Review of NASA'S Distributed Active Archive Centers

Committee on Geophysical and Environmental Data
Board on Earth Sciences and Resources
Commission on Geosciences, Environment, and Resources
National Research Council

NATIONAL ACADEMY PRESS
Washington, D.C 1998

NATIONAL ACADEMY PRESS • 2101 Constitution Avenue, N.W. • Washington, DC 20418

NOTICE: The project that is the subject of this report was approved by the Governing Board of the National Research Council, whose members are drawn from the councils of the National Academy of Sciences, the National Academy of Engineering, and the Institute of Medicine. The members of the committee responsible for the report were chosen for their special competences and with regard for appropriate balance.

Support for this study was provided by Award No. NCC5-202 between the National Academy of Sciences and the National Aeronautics and Space Administration.

International Standard Book Number 0-309-06331-0

Additional copies of this report are available from:

National Academy Press
2101 Constitution Avenue, NW
Box 285
Washington, D.C. 20055
800-624-6242
202-334-3313 (in the Washington Metropolitan Area)
http://www.nap.edu

Cover: Collage of images from the Tropical Rainfall Measuring Mission (TRMM). The globes show probable rainfall as indicated by the TRMM Microwave Imager. TRMM's precipitation radar gathered data on the amount and distribution of rainfall and clouds in Hurricane Bonnie during August 1998. The colors of the profile represent surface rainfall rates varying from 0 mm/hr (dark blue) to 50 mm/hr (dark red). Images courtesy of C. Morales, G. Shirah, R. Simmon, and M. Sutton, NASA GSFC.

Printed in the United States of America

COMMITTEE ON GEOPHYSICAL AND ENVIRONMENTAL DATA

FRANCIS BRETHERTON, *Chair*, University of Wisconsin, Madison
VERA ALEXANDER, University of Alaska, Fairbanks (through 1997)
DAVID GLOVER, Woods Hole Oceanographic Institution, Massachusetts
MALCOLM K. HUGHES, University of Arizona, Tucson (through 1997)
J.-BERNARD MINSTER, Scripps Institution of Oceanography, La Jolla, California
WILLIAM J. PARTON, JR., Colorado State University, Fort Collins
JOYCE E. PENNER, University of Michigan, Ann Arbor
THOMAS A. POTEMRA, The Johns Hopkins University, Laurel, Maryland (through 1997)
SOROOSH SOROOSHIAN, University of Arizona, Tucson
ELLEN MOSLEY-THOMPSON, The Ohio State University, Columbus (through 1997)
JOHN R. G. TOWNSHEND, University of Maryland, College Park
CARL WUNSCH, Massachusetts Institute of Technology, Cambridge

National Research Council Staff

ANNE M. LINN, Study Director
ANTHONY R. DE SOUZA, Senior Program Officer
JENNIFER T. ESTEP, Administrative Assistant

PANELS TO REVIEW NASA'S DISTRIBUTED ACTIVE ARCHIVE CENTERS

Alaska SAR Facility

J.-BERNARD MINSTER, *Chair*, Scripps Institution of Oceanography, La Jolla, California
FERRIS WEBSTER, *Vice Chair*, University of Delaware, Lewes
KURT L. FEIGL, Groupe de Recherche de Geodesie Spatiale, Toulouse, France
ANTHONY C. HEARN, RAND Corporation, Santa Monica, California
MICHAEL S. LOUGHRIDGE, NOAA National Geophysical Data Center, Boulder, Colorado
JOHN M. MELACK, University of California, Santa Barbara
ROBERT S. PRITCHARD, IceCasting, Inc., San Rafael, California

EROS Data Center

J.-BERNARD MINSTER, *Chair*, Scripps Institution of Oceanography, La Jolla, California
FERRIS WEBSTER, *Vice Chair*, University of Delaware, Lewes
RUTH S. DEFRIES, University of Maryland, College Park
ALLEN M. HITTELMAN, NOAA National Geophysical Data Center, Boulder, Colorado
DAVID A. LANDGREBE, Purdue University, West Lafayette, Indiana
DAR ROBERTS, University of California, Santa Barbara

Goddard Space Flight Center

J.-BERNARD MINSTER, *Chair*, Scripps Institution of Oceanography, La Jolla, California
FERRIS WEBSTER, *Vice Chair*, University of Delaware, Lewes
SYDNEY LEVITUS, NOAA National Oceanographic Data Center, Silver Spring, Maryland
RICHARD S. LINDZEN, Massachusetts Institute of Technology, Cambridge
TERENCE R. SMITH, University of California, Santa Barbara
JOHN R. G. TOWNSHEND, University of Maryland, College Park

Jet Propulsion Laboratory (PO.DAAC)

J.-BERNARD MINSTER, *Chair*, Scripps Institution of Oceanography, La Jolla, California
FERRIS WEBSTER, *Vice Chair*, University of Delaware, Lewes
DAVID DEWITT, University of Wisconsin, Madison
SYDNEY LEVITUS, NOAA National Oceanographic Data Center, Silver Spring, Maryland
LYNNE D. TALLEY, Scripps Institution of Oceanography, La Jolla, California

Langley Research Center

J.-BERNARD MINSTER, *Chair*, Scripps Institution of Oceanography, La Jolla, California
FERRIS WEBSTER, *Vice Chair*, University of Delaware, Lewes
C. BRUCE BAKER, NOAA National Climatic Data Center, Asheville, North Carolina
ROBERT D. CESS, State University of New York, Stony Brook
RUSSELL R. DICKERSON, University of Maryland, College Park
REX W. TRACY, GDE Systems, Inc., San Diego, California

National Snow and Ice Data Center

J.-BERNARD MINSTER, *Chair*, Scripps Institution of Oceanography, La Jolla, California
FERRIS WEBSTER, *Vice Chair*, University of Delaware, Lewes
VERA ALEXANDER, University of Alaska, Fairbanks
KEITH E. LITTLEFIELD, National Imagery and Mapping Agency, Fairfax, Virginia
JOHN E. WALSH, University of Illinois, Urbana
J. RONALD WILSON, Marine Environmental Data Service (retired), Manotick, Ontario

Oak Ridge National Laboratory

J.-BERNARD MINSTER, *Chair*, Scripps Institution of Oceanography, La Jolla, California
FERRIS WEBSTER, *Vice Chair*, University of Delaware, Lewes
KENNETH D. DAVIDSON, NOAA National Climatic Data Center, Asheville, North Carolina
BETH H. DRIVER, National Imagery and Mapping Agency, Chantilly, Virginia
WILLIAM J. PARTON, JR., Colorado State University, Fort Collins
ERIC T. SUNDQUIST, U.S. Geological Survey, Woods Hole, Massachusetts

National Research Council Staff

ANNE M. LINN, Study Director
JENNIFER T. ESTEP, Administrative Assistant

BOARD ON EARTH SCIENCES AND RESOURCES

COMMISSION ON GEOSCIENCES, ENVIRONMENT, AND RESOURCES

The National Academy of Sciences is a private, nonprofit, self-perpetuating society of distinguished scholars engaged in scientific and engineering research, dedicated to the furtherance of science and technology and to their use for the general welfare. Upon the authority of the charter granted to it by the Congress in 1863, the Academy has a mandate that requires it to advise the federal government on scientific and technical matters. Dr. Bruce Alberts is president of the National Academy of Sciences.

The National Academy of Engineering was established in 1964, under the charter of the National Academy of Sciences, as a parallel organization of outstanding engineers. It is autonomous in its administration and in the selection of its members, sharing with the National Academy of Sciences the responsibility for advising the federal government. The National Academy of Engineering also sponsors engineering programs aimed at meeting national needs, encourages education and research, and recognizes the superior achievements of engineers. Dr. William A. Wulf is president of the National Academy of Engineering.

The Institute of Medicine was established in 1970 by the National Academy of Sciences to secure the services of eminent members of appropriate professions in the examination of policy matters pertaining to the health of the public. The Institute acts under the responsibility given to the National Academy of Sciences by its congressional charter to be an adviser to the federal government and, upon its own initiative, to identify issues of medical care, research, and education. Dr. Kenneth I. Shine is president of the Institute of Medicine.

The National Research Council was organized by the National Academy of Sciences in 1916 to associate the broad community of science and technology with the Academy's purposes of furthering knowledge and advising the federal government. Functioning in accordance with general policies determined by the Academy, the Council has become the principal operating agency of both the National Academy of Sciences and the National Academy of Engineering in providing services to the government, the public, and the scientific and engineering communities. The Council is administered jointly by both Academies and the Institute of Medicine. Dr. Bruce Alberts and Dr. William A. Wulf are chairman and vice-chairman, respectively, of the National Research Council.

Acknowledgment of Reviewers

This report has been reviewed by individuals chosen for their diverse perspectives and technical expertise in accordance with procedures approved by the NRC's Report Review Committee. The purpose of this independent review is to provide candid and critical comments that will assist the authors and the NRC in making their published report as sound as possible and to ensure that the report meets institutional standards for objectivity, evidence, and responsiveness to the study charge. The content of the review comments and draft manuscript remain confidential to protect the integrity of the deliberative process. We wish to thank the following individuals for their participation in the review of this report: Fakhri A. Bazzaz, Harvard University; Ingrid C. Burke, Colorado State University; William L. Chameides, Georgia Institute of Technology; Charles S. Cox, Scripps Institution of Oceanography; Robert E. Dickinson, University of Arizona; Timothy H. Dixon, University of Miami; Jeff Dozier, University of California, Santa Barbara; Edward A. Frieman, Scripps Institution of Oceanography; Robert A. Frosch, Harvard University; Eric S. Kasischke, ERIM; Thomas F. Malone, North Carolina State University (*retired*); Ichtiaque Rasool, International Consultant; Robert H. Stewart, Texas A&M University; Edwin D. Waddington, University of Washington; Wilford F. Weeks, University of Alaska (*emeritus*); Carol Wessman, University of Colorado; and Mary Lou Zoback, U.S. Geological Survey.

Although the individuals listed above have provided many constructive comments and suggestions, responsibility for the final content of this report rests solely with the authoring committee and the NRC.

Preface

The long-term health of data centers depends on their ability to keep pace with technological advances that improve user services and increase the long-term utility of the data, and to respond to the evolving needs of their user communities. Consequently, it is important for data centers to be reviewed externally on a periodic basis. This report, which was requested by Robert Price, then associate director of Goddard for Mission to Planet Earth (now the Earth Science Enterprise), is the first such assessment of seven Distributed Active Archive Centers (DAACs). The DAACs, which manage a wide variety of satellite and *in situ* measurements associated with the National Aeronautics and Space Administration's (NASA's) Earth Observing System Data and Information System (EOSDIS), are undergoing a recertification process with the overall goal of improving their operations. NASA's recertification process will draw upon several inputs, including the results of external peer review, which are reported here. At the end of the process, NASA management will determine whether to recertify, place on probation, or close individual DAACs.

The National Research Council's (NRC's) Committee on Geophysical and Environmental Data (CGED) approached the review process as it has for other data center reviews since it was assigned oversight of U.S. World Data Centers in 1967. The simultaneous review of seven DAACs is a daunting task, so the CGED divided the review into two stages. In the first stage, two CGED members visited each DAAC informally to identify key issues. Based on these visits, the committee defined review criteria, which focus on how well the DAACs serve their scientific user communities. The formal site visits, which were conducted by seven separately appointed panels, composed the second stage of the review. The formal site visits were held on the following dates:

DAAC Host Institution	*Scientific Specialty*	*Site Visit Date*
Goddard Space Flight Center	Atmospheric processes	October 20-21, 1997
Langley Research Center	Atmospheric chemistry	November 18-19, 1997
EROS Data Center	Land processes	November 24-25, 1997
Alaska SAR Facility	Sea ice, polar processes	December 18-19, 1997
Jet Propulsion Laboratory	Ocean circulation	January 8-9, 1998
National Snow and Ice Data Center	Snow and ice, cryosphere	March 4-5, 1998
Oak Ridge National Laboratory	Biogeochemical processes	March 19-20, 1998

At NASA's request, the DAAC located at the Consortium for Integrated Earth Science Information Network (CIESIN) was not reviewed.

Each site visit panel was composed of approximately six individuals. For consistency, the chair and vice chair were the same for each visit. Those individuals were chosen for their links with the CGED and their familiarity with data center reviews. The rest of the panel was composed of

- two scientists, who use or collect the types of data held by the DAAC;
- one representative of a non-NASA scientific data center, familiar with data center operations; and
- one computer scientist or systems engineer, with experience in end-to-end system management and a knowledge of networks, computer architecture, and the types and capability of computer equipment available for managing large data sets.

In addition, David Glover, who chaired NASA's EOSDIS Panel at the time of the review, attended all the site visits as an observer.

For each review, the panel followed the same agenda and received similar briefing materials from the DAAC. The briefing materials included the DAAC's annual work plan, including staffing and budget projections; membership list and meeting minutes of the DAAC's User Working Group; and in most cases, written responses to the criteria for review. In addition, two members of each panel examined the DAAC's Web site, accessed data of interest to their own research, and evaluated issues such as documentation, formats, ease of use, and the ability of User Services to answer questions. As a result, the site visits were conducted at approximately the same level of detail, yielding a fair and balanced picture of the DAACs. The individual DAAC reports, however, vary significantly in emphasis, which is unavoidable given the different backgrounds and personalities of the authoring panels.

To place the panels' reports within the broader EOSDIS context, the CGED solicited input from a variety of sources, including the Earth Science Data and Information System (ESDIS) Project, the EOSDIS Core System (ECS) developers, and users of EOSDIS data. A two-day interview with ESDIS Project managers focused on (1) ESDIS expectations of the DAACs; (2) DAAC expectations of ESDIS, based on issues identified during the informal CGED visits to the DAACs; and (3) other issues raised by previous NRC reports, particularly the Zraket panel report (NRC, 1994). Interviews with the ECS developers (and Jet Propulsion Laboratory [JPL] developers for the Alaska Synthetic Aperture Radar [SAR] Facility DAAC) were conducted by subpanels, which included the chair and vice chair of the panels and a computer scientist. Finally, an e-mail survey of users was sent to the EOS Investigators Working Group and subsequently forwarded to a broader audience. Nearly 400 users responded. Rigorous statistical analysis of this unscientific survey is not justified, but general trends and comments from individuals helped illustrate the broad range of experiences that users have had with the DAACs. The results of these interviews, the user survey, and the panel reports formed the basis for the overall conclusions and recommendations of this report.

It is important to note that plans for EOSDIS evolved significantly during the course of the CGED review. For example, the launch data of the EOS AM-1 platform slipped by at least six months, giving the DAACs more time to prepare for the data streams. On the other hand, additional delays in the ECS are causing the DAACs and EOS science and instrument teams to resort to emergency backup plans for processing the data. These plans are evolving on almost a daily basis. This report attempts to provide a snapshot of the DAAC system as it existed in September 1998. The panel reports were updated through e-mail correspondence with the DAAC managers and by an additional site visit to the first DAAC visited (Goddard Space Flight Center). Similarly, the overview chapters were written to account for recent developments in EOSDIS. The committee believes that the resulting report can be used as a baseline for future assessments of the health of the DAAC system or for more focused reviews of individual DAACs.

The committee and its panels wish to acknowledge the many individuals who provided input to this report. They include the current and past DAAC managers (Paul Chan, Donald Collins, Roy Dunkum, Craig Lingle, Richard McGinnis, Lydon Oleson, Larry Voorhees, Carl Wales, and Ron Weaver) and staff, whose forthright participation in the review permitted the panels to examine weaknesses as well as strengths of DAAC operations. Representatives of the DAACs' host institutions helped the panels identify distinctions between DAACs located within a university and those located within a NASA or non-NASA government facility. Discussions with members of the User Working Groups, science teams, system developers, and ESDIS management provided the panels with a more complete view of the DAACs as part of an integrated EOSDIS. The nearly 400 survey respondents from the United States and abroad gave the committee a better sense of users' overall satisfaction with the DAACs and patterns of DAAC usage.

The committee and its panels thank the staff of the Board on Earth Sciences and Resources of the National Research Council, especially Jenny Estep, for helping very effectively with the unusually complex logistics of innumerable meetings and site visits. Finally, we wish to express particular appreciation to the indefatigable study director, Anne Linn, for her invaluable guidance through the review process, and for her very hard work during the production of this report. The combination of her superb sense of organization, insight, determination, and patience contributed the necessary ingredients for the successful completion of this daunting task.

Francis Bretherton
Chair, CGED

Bernard Minster
Chair, site visit panels

Contents

EXECUTIVE SUMMARY **1**

1 **INTRODUCTION** **9**
Background, 9
Role of the DAACs in EOSDIS, 11
Evolution of EOSDIS, 20
Organization of Report, 24

2 **OVERVIEW OF THE DAAC SYSTEM** **27**
What Is a DAAC? Why a System?, 27
The Need for Adaptability, 32
DAACs and Their Users, 34
Data Management: Soup to Nuts, 41
The Role of NASA, 47
Other Issues, 49
Conclusions, 52

3 **GODDARD SPACE FLIGHT CENTER DAAC** **53**
Abstract, 53
Introduction, 54
Holdings, 54
Users, 60
Technology, 62
Management, 64
GSFC DAAC and the Earth Science Enterprise, 67
Summary, 68

4 LANGLEY RESEARCH CENTER DAAC **71**

Abstract, 71
Introduction, 72
Holdings, 72
Users, 79
Technology, 82
Management, 85
LARC DAAC and the Earth Science Enterprise, 87
Summary, 88

5 EROS DATA CENTER DAAC **91**

Abstract, 91
Introduction, 92
Holdings, 93
Users, 97
Technology, 101
Management, 103
EDC DAAC and the Earth Science Enterprise, 105
Summary, 106

6 ALASKA SAR FACILITY DAAC **109**

Abstract, 109
Introduction, 110
Holdings, 111
Users, 116
Technology, 120
Management, 123
ASF DAAC and the Earth Science Enterprise, 125
Summary, 128

7 PHYSICAL OCEANOGRAPHY DAAC **131**

Abstract, 131
Introduction, 132
Holdings, 133
Users, 136
Technology, 138
Management, 140
PO.DAAC and the Earth Science Enterprise, 143
Summary, 144

8 NATIONAL SNOW AND ICE DATA CENTER DAAC **147**
Abstract, 147
Introduction, 148
Holdings, 149
Users, 152
Technology, 156
Management, 158
NSIDC DAAC and the Earth Science Enterprise, 161
Summary, 163

9 OAK RIDGE NATIONAL LABORATORY DAAC **165**
Abstract, 165
Introduction, 166
Holdings, 167
Users, 172
Technology, 174
Management, 175
ORNL DAAC and the Earth Science Enterprise, 179
Summary, 180

AFTERWORD **183**

REFERENCES **185**

APPENDIXES **187**
Appendix A DAAC Review Agenda, 189
Appendix B Criteria for Review, 191
Appendix C CGED Interview with ESDIS (September 1997), 195
Appendix D User Survey, 215
Acronyms, 231

Executive Summary

The National Aeronautics and Space Administration's (NASA's) Distributed Active Archive Centers (DAACs) are a key part of the Earth Observing System Data and Information System (EOSDIS). Each DAAC has an essential and individual role in some part of the scientific enterprise, with little overlap or redundancy. Given the uncertainties associated with the EOSDIS Core System, those affected by the near-term launch of the EOS AM-1 platform (a satellite with a 10:00 a.m. sun-synchronous orbit and a large number of instruments) are reasonably placed to address the challenges entailed by this launch. However, for the DAACs effectively to fulfill the expectations for them, EOSDIS will need inspired leadership to create a practical network of information centers that truly enables scientific discovery and assessment and integrates the creative energies of the DAACs, their scientific communities, and the pilot Federation of Earth System Information Partners.

INTRODUCTION

The objective of NASA's Earth Science Enterprise (ESE) is to obtain a broad-based understanding of the functioning of the Earth as a system, with particular reference to global environmental change. The data needed to study these processes are necessarily diverse, comprising observations from a wide variety of remote sensing and *in situ* instruments and experiments, collected by several government agencies and academic institutions, at different temporal and spatial scales. NASA's Earth Observing System Data and Information System was built to provide a means for scientists to integrate disparate data types collected by NASA and to study earth processes in a more comprehensive manner than was possible before.

EOSDIS includes many players—the EOSDIS Core System (ECS) contractor, the science and instrument teams, the DAACs, and the Earth Science Data and Information System (ESDIS) Project—each with well-defined roles. As originally conceived, these roles were as follows:

- the ECS contractor designs the EOSDIS Core System to capture, process, and distribute data from the EOS instruments and provides the necessary hardware and software to the DAACs;
- the science and instrument teams develop algorithms for creating data products;
- the seven DAACs process and disseminate remote sensing and *in situ* data and data products (land, atmosphere, ocean), and provide services to a wide variety of users (primarily scientists and NASA's partner agencies); and
- NASA's ESDIS Project sets the program requirements and provides funding and system-wide coordination.

Technical problems leading to delays in the ECS, however, have led NASA to rethink these roles. NASA's current plans are to make EOSDIS a more distributed system by having the science and instrument teams, rather than the DAACs, do much of the data processing. In addition, some of the DAACs will be permitted to develop and use their own systems, rather than the ECS, for managing data. Indeed, DAAC-unique information systems are already being used to process and distribute data from the first EOS-related mission, the Tropical Rainfall Measuring Mission (TRMM), which was launched in November 1997. By involving a broader array of constituencies in EOSDIS, and employing diverse approaches to providing data services, NASA is taking another step toward creating an EOSDIS federation.

In the new EOSDIS model, the role of the DAACs may be strengthened or diminished, depending on how much flexibility and authority NASA management is prepared to give and how much initiative the DAACs are prepared to take. However, serving the needs of their users will remain their most important task. Based on the site visit reports of the seven DAAC review panels and a user survey, the Committee on Geophysical and Environmental Data (CGED) concludes that most of the DAACs are serving their scientific user community well. (Indeed, the Physical Oceanography [PO.DAAC] and the National Snow and Ice Data Center [NSIDC] DAACs are model in this regard, although neither has to face the immediate challenge of handling the enormous data streams of the AM-

1 platform.) Moreover, each DAAC occupies a unique, scientifically important niche in the Earth Science Enterprise; closing any of them would reduce NASA's ability to meet its scientific objectives. Consequently, even DAACs with significant problems in fulfilling their missions (i.e., the Alaska Synthetic Aperture Radar [SAR] Facility [ASF] DAAC and, to a lesser extent, the Earth Resources Observation Systems [EROS] Data Center [EDC] and the Oak Ridge National Laboratory [ORNL] DAACs) should be nurtured by NASA and the ESDIS Project so that they succeed.

The individual panel reports provide detailed recommendations on improving the operations of each DAAC. However, all of the reports have three themes in common: (1) the scientific need for a coherent system of DAACs; (2) the importance of strategic planning in routine data center operations; and (3) the need for flexibility, vision, and leadership as EOSDIS evolves. The following recommendations, which focus on these themes, were based on analysis of the panel reports, the user survey, discussions with NASA managers, and the committee's experience with world and national data center reviews.

AN INFORMATION SYSTEM FOR SCIENCE

EOSDIS is more than the hardware and software needed to link the DAACs; it is a tool for achieving the science goals of the Earth Science Enterprise. The objectives of EOSDIS include the following:

- facilitating the creation of standard data products, thereby permitting the immediate scientific goals of the science teams to be realized;
- catalyzing the preparation of a wide range of secondary data sets and information products that combine information from different satellites and *in situ* sources, thereby stimulating collaborative, multidisciplinary research;
- making such products readily accessible to the broader scientific community; and
- preserving them in usable form for future generations of scientists.

The DAACs have a key role to play in meeting each of these objectives. First, by working with the science and instrument teams (as the Langley Research Center [LaRC] DAAC has on Earth Radiation Budget experiments), the DAACs can help generate data products and record metadata that might be viewed as irrelevant or common knowledge by science and instrument team members. The latter is vital if the data and data products are to retain their scientific value over several decades. Second, by working with the broader scientific user community, the DAACs can build a mutually beneficial relationship that results in (1) the development of useful tools and services, (2) the acquisition of scientifically important data sets, and (3) better answers to complex user queries. In practice, however, only a few DAACs (namely the PO.DAAC and the NSIDC DAAC) have a strong relationship with the scientific community, despite their generally good interaction with User Working Groups and their collocation with scientists with expertise in the DAAC holdings.

Recommendation. To function optimally, the DAACs need to be intimately involved with the scientific community they serve. The DAACs should deliberately pursue and improve routine, daily interactions with active scientists who use their data holdings. Among the many ways to meet this recommendation, the committee suggests (1) implementing a visiting scientist program; (2) encouraging DAAC personnel to pursue research endeavors, with the purpose of publishing the results; and (3) actively working with researchers within their host institution.

Multidisciplinary researchers interested in understanding earth system science processes present a special challenge to the DAACs. These researchers need to obtain useful data from more than one DAAC. Indeed, it was to serve their needs that the concept of "one-stop shopping" (i.e., users can access the system through any DAAC, search all the EOSDIS holdings, and obtain the relevant data) was born. With the delays in the ECS and the move toward federation, however, one-stop shopping may no longer be feasible. For EOSDIS to serve the earth system science constituency, the DAACs will have to overcome their differences and begin to act as components of a coherent system that (1) is able to manage high-volume data streams from coordinated instruments and (2) offers comparable access technologies, a consistent terminology, and unobtrusive authorization procedures. If such a system is not possible, NASA will have to provide sufficient resources (e.g., through requests for proposals to participate in NASA's federation) to enable users to find useful information from a disparate collection of DAACs. The development of Version 0 of the EOSDIS information system demonstrated the ability of the DAACs to work together. With the proper incentives from ESDIS, the DAACs should also be able to create common tools and standards that would enable users to locate, access, and combine various types of data in the EOS era.

Recommendation. The DAACs do not yet act as components of a coherent system. They share the responsibility for providing the vision and leadership toward this goal with the science teams, ESDIS, and NASA. If such a coherent system cannot be achieved, NASA should place a greater emphasis on user-generated proposals seeking to help the community deal with a disparate DAAC system.

Recommendation. A DAAC alliance with a common goal will better serve the broader community than the collection of individual centers that currently exists. The DAACs should support each other and express a collective point of view on EOSDIS policies.

STRATEGIC PLANNING

DAACs differ from data centers by focusing on the early, most scientifically active part of the mission or experiment, rather than on long-term data stewardship. To fulfill the science objectives of the Earth Science Enterprise, however, the DAACs also have an obligation to ensure that the data in their charge remain useful for future generations of scientists studying long-term global environmental change. Consequently, the DAACs must be involved in all stages of data management—from data collection, to management of active data sets, to long-term archive. Such involvement requires strategic planning.

To date, however, few DAACs have engaged in strategic planning, in part because ESDIS and the ECS contractor have done so much of their planning for them. (The PO.DAAC is an exception.) However, now that NASA is adopting some of the DAAC fallback systems for processing and handling data in place of the ECS, the DAACs will have a greater voice in their own evolution and in the development of EOSDIS. To take advantage of this opportunity, each DAAC should create a vision and an implementation strategy. The strategy should permeate everything the DAAC does. For example, it should describe how the DAACs will become involved in collecting the metadata and in calibrating and validating observations from the EOS instruments. The latter is particularly important for the ORNL DAAC, which has important *in situ* data sets, but no concrete plans for participating in calibration or validation activities with the relevant science teams. Without the integration of remote sensing and *in situ* observations, it is unlikely that the EOS program will ever fulfill its potential.

The vision and implementation strategy should also describe how the DAACs manage their active data sets. This is a primary focus of the DAACs and includes knowing who their current users are, how they use the data, what tools are available for serving them, and which ones must be developed in house. It also includes looking ahead to identify and provide services to potential new user communities and to keep up with rapid advances in technology, particularly those involving storage and communications. Most of the DAACs manage their active data sets well and provide a reasonably high level of service to users. The PO.DAAC, the ORNL DAAC, and the NSIDC DAAC are particularly effective in this regard, whereas the ASF and EDC DAACs have to substantially improve user services. All the DAACs, however, would serve their users better by (1) having a more detailed understanding of their user profile and (2) developing quantitative measures for tracking performance. A detailed user profile, monitored over time, will enable the DAACs to provide specialized services to current and potential user communities and to expand their user base. Promoting new uses of the data, which will arise inevitably with the emergence of new user groups, will increase the taxpayer's return on investment in the EOS satellites. In addition, by developing and tracking quantitative performance measures, the DAACs will be able to assess the technical performance of individual system components, as well as the overall success of each DAAC and the entire DAAC system in meeting the needs of their users.

Recommendation. In order better to track the rapidly growing and evolving population of EOSDIS users and serve the needs of existing and potential users, each DAAC should devise and implement quantitative measures for characterizing its current user community.

Recommendation. Ongoing changes in data volumes, user expectations, and emerging technologies are powerful forces that put pressure on each DAAC to evolve independently of the others. In order to counteract such centrifugal forces, each DAAC should prepare and periodically update a practical strategic plan for dealing with change, while preserving the concept of a coherent system.

Recommendation. Excellence in a research enterprise is best gauged through assessment of performance by one's peers, according to a commonly accepted set of performance criteria. The DAACs must develop a set of quantitative performance metrics by which they can measure their own progress and evaluate their success as individual centers and as a coherent system. Periodic peer review aimed at gauging accomplishments against these metrics should be incorporated as part of this ongoing process.

FLEXIBILITY IN EOSDIS

With the delays in the ECS, NASA has no choice now but to abandon many components of the ECS and move toward a more flexible federated structure for EOSDIS. Changing management models at this late stage poses a tremendous challenge to NASA. For the new EOSDIS paradigm to succeed, NASA must (1) empower the DAACs, science teams, and Earth Science Information Partners (ESIPs) to provide the best services and most innovative data products possible; (2) offer incentives for them to collaborate to provide users with a common look and feel to the information system; and (3) emplace safeguards to ensure that the fundamental goals of EOSDIS are met. To accomplish the first, ESDIS will have to delegate some of its authority for serving users and designing their own information systems to the DAACs. It has done so to some extent with the Goddard Space Flight Center and LaRC DAACs, which successfully developed TRMM-specific information systems. Similar authority should be granted immediately to the ASF DAAC, which is currently powerless to guide development of the information system it is required to use (which was designed and controlled by the Jet Propulsion Laboratory). In developing their information systems, some of the DAACs may wish to use parts or all of the ECS, but this should not be a requirement for remaining within EOSDIS. In return for this newfound authority and flexibility, the DAACs must agree to act as components of a coherent system for the greater good of the Earth Science Enterprise (see "*An Information System for Science*," above). This is particularly important for the LaRC DAAC, which has no intention of linking its custom-built systems with EOSDIS.

Recommendation. The DAAC-ECS-ESDIS model for managing EOSDIS data and information has not succeeded. To take advantage of new technological approaches and management models, ESDIS should foster the creation of a federation of DAACs by delegating to the DAACs some of its authority for serving users and by providing incentives to the DAACs to serve the broader community as well as their narrow constituencies.

Recommendation. To take advantage of the unprecedented flexibility afforded by the new Web-based technologies, ESDIS should allow the DAACs to incorporate only those components of the ECS that they require to satisfy their user community. This flexibility should not come at the cost of reducing the DAACs' ability to function as full-fledged members of the DAAC system. For the DAACs, the price of this flexibility is an increased individual responsibility to contribute to the overall goals of EOSDIS.

In summary, every DAAC has a unique and important role in some part of EOSDIS and the Earth Science Enterprise, with little overlap or redundancy. The DAACs that will receive data from the AM-1 platform and other near-term launches are reasonably placed to address the challenges that launch will entail, given the uncertainties associated with the EOSDIS Core System. However, for the DAACs to fulfill their mission of providing seamless access to multidisciplinary data sets, and thereby promoting creative scientific analysis of the data, EOSDIS will need inspired leadership that empowers its stakeholders to fulfill their special roles within an integrated network of information centers.

1

Introduction

BACKGROUND

There can be no question that the National Aeronautics and Space Administration's (NASA's) Earth Observing System (EOS)—and its implementation into the Mission to Planet Earth (MTPE), now renamed the Earth Science Enterprise (ESE)—is the most ambitious effort ever attempted to study our planet and its set of complex interlocking systems. Although the original concept has evolved substantially over the years, the overall scientific goal endures, and "earth system science" (see NASA, 1988) has gained the respectability of a *bona fide* discipline.

One of the major components of the enterprise has been, and remains to this day, the EOS Data and Information System (EOSDIS). EOSDIS itself is an ambitious endeavor, not only because it must deal with data volumes and acquisition rates that are several orders of magnitude larger than ever before in the earth sciences, but also because its stated goal is to permit seamless access to multidisciplinary data in a timely manner. This has never been attempted before on such a scale, and the hope is that EOSDIS will facilitate research that would not be contemplated otherwise. Providing access to EOSDIS data is the job of the Distributed Active Archive Centers (DAACs). Although the DAAC system *per se* is not the largest component of EOSDIS in terms of cost, it holds a critical position in the overall architecture of the system, first because the user interacts with the DAACs and second because the burden of maintaining operational access to large and complex multidisciplinary data sets befalls them most directly.

Ever since they were chosen by NASA, the DAACs have drawn both praise and criticism for their ability to serve their users. In some cases, the DAACs were criticized by implication as a result of reviews of EOS or EOSDIS. However, the DAACs have never been systematically assessed as components of a comprehensive data and information system. NASA has initiated such an assessment through a "recertification" process, which is being conducted in two stages. The first stage, an external peer review of the DAACs, has been conducted under the auspices of the National Research Council's (NRC's) Committee on Geophysical and Environmental Data (CGED). This report gives the results of that peer review, the first that the DAACs have ever undergone. The second stage of the recertification process will be conducted by a NASA panel, which will evaluate the results of the peer review in the context of NASA's programmatic and budgetary priorities. NASA management will then decide whether to recertify, place on probation, or close individual DAACs.

Each DAAC manages a different kind of scientific data—atmospheric, oceanic, solid-earth, polar, biospheric—and serves a unique blend of user communities. Because no single committee has the appropriate composition to review all the centers, seven review panels were established to conduct the site visits. The DAACs reviewed are located at the Goddard Space Flight Center (GSFC), Langley Research Center (LaRC), Earth Resources Observation Systems (EROS) Data Center (EDC), Alaska Synthetic Aperture Radar (SAR) Facility (ASF), Jet Propulsion Laboratory (JPL; Physical Oceanography DAAC [PO.DAAC]), National Snow and Ice Data Center (NSIDC), and the Oak Ridge National Laboratory (ORNL). (At NASA's request, the Socio-Economic DAAC [SEDAC] located at the Consortium for International Earth Science Information Networks [CIESIN] was not reviewed.) The CGED established the criteria for review and provided the panels with similar agendas and briefing materials so that the site visits would be conducted as uniformly as possible. A description of the study process is given in the Preface to this report.

In this report the CGED does not recommend whether or not any DAACs should be closed or placed on probation. Rather, based on the criteria for review listed in Appendix B, the committee and its panels commend the DAACs' successes and identify issues that require greater attention, with the overall goal of improving the DAACs' ability to serve their users. EOS is a science program that is designed to serve scientists. Thus, the CGED review focuses primarily on how well the DAACs serve the scientific community and secondarily on other types of users. Finally, because the DAACs exist as part of a system rather than as independent entities, the committee also addresses overarching issues regarding the DAACs as components of EOSDIS. (In keeping with its charge, the committee evaluated the DAAC's performance against their mission, but not against alternative ways of achieving the same goals. Similarly, the committee did not review EOSDIS as a whole, or the role of EOSDIS in global change research. The latter is addressed in NRC, 1998a.) This report was written at a time when technical difficulties, budgetary pressures, technological advances, and new management approaches were fast changing the face of EOSDIS. Nonetheless, the committee believes that the report provides a baseline from which the progress of individual DAACs or the DAAC system as a whole can be measured.

ROLE OF THE DAACS IN EOSDIS

EOSDIS was designed to perform a variety of functions, from spacecraft command and control, to data acquisition, processing, distribution, and archive. Linking these disparate functions are the hardware and software that comprise the EOSDIS Core System (ECS). The architecture of EOSDIS, as defined by NASA, is illustrated schematically in Figure 1.1.

In general, data acquired from spacecraft will be captured and processed by the ECS contractor to Level 0 (see Table 1.1), then transferred to the DAACs. Some DAACs also receive *in situ* data, usually from principal investigators of field experiments or process studies. Such data will be used to calibrate and validate the satellite measurements and gain a more complete understanding of the phenomena being studied. Using algorithms developed by the science and instrument teams, the DAACs and/or science teams will process the Level 0 data into Level 1 and higher standard data products. Some of the data will be processed in near real time, although this is not an EOSDIS requirement, given the complexity of many of the algorithms. Typically, the higher-level products will be processed by the science teams. The data products will then be disseminated by the DAACs, which will also provide support services to users and archive the data and data products.

The DAAC System

The DAACs' partners in EOSDIS include the ECS contractor, science and instrument teams, and the Earth Science Data and Information System (ESDIS) Project. Their primary roles, which are discussed in more detail below, are as follows:

- the ECS contractor builds the information system to capture and process data, and link the DAACs together;
- instrument and science teams develop algorithms for processing data and generate data products;
- DAACs process and disseminate data and provide user services; and
- the ESDIS Project sets the requirements for the information system and coordinates the DAAC system.

An organization chart, with the reporting lines within EOSDIS is shown in Figure 1.2.

The DAACs themselves form a loose consortium for solving problems in

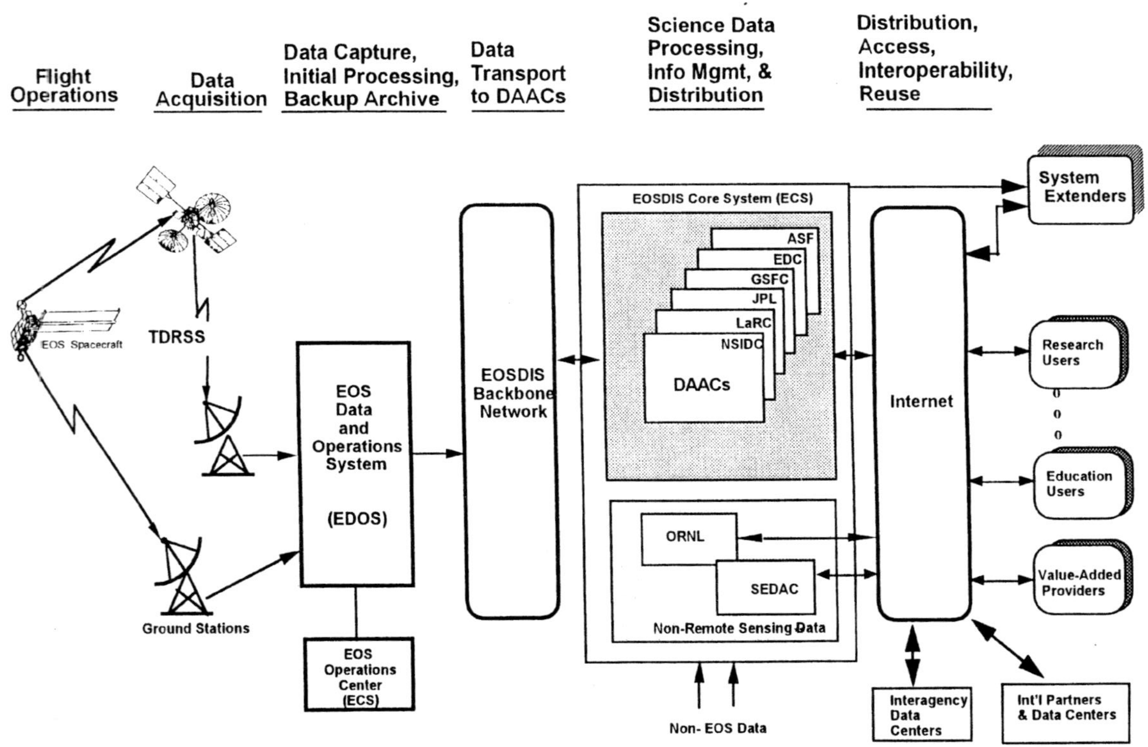

FIGURE 1.1. EOSDIS architecture, from flight operations and data acquisition (TDRSS = Tracking and Data Relay Satellite System) to data processing and dissemination. NASA has divided the DAACs into centers handling remote sensing data (ASF, EDC, GSFC, JPL [PO.DAAC], LaRC, and NSIDC DAACs) and centers handling *in situ* and other types of data (ORNL DAAC and SEDAC). SOURCE: NASA Headquarters.

TABLE 1.1. Data Set Processing Levels

Data Level	Description
Level 0	Reconstructed unprocessed instrument or payload data at full resolution; any and all communications artifacts (e.g., synchronization frames, communications headers) removed
Level 1A	Reconstructed unprocessed instrument data at full resolution, time referenced, and annotated with ancillary information, including radiometric and geometric calibration coefficients and georeferencing parameters (i.e., platform ephemeris) computed and appended, but not applied, to the Level 0 data
Level 1B	Level 1A data that have been processed to sensor units (not all instruments will have a Level 1B equivalent)
Level 1C	TRMM-specific for quality content of Level 1B precipitation radar and ground validation data
Level 2	Derived geophysical variables at the same resolution and location as the Level 1 source data
Level 3	Variables mapped on uniform space-time grid scales, usually with some completeness and consistency
Level 4	Model output or results from analyses of lower-level data (i.e., variables derived from multiple measurements)

NOTE: TRMM = Tropical Rainfall Measuring Mission.
SOURCE: Asrar and Greenstone (1995).

common. For example, the DAACs worked collectively to implement Version 0 of EOSDIS, which permits users to browse the holdings of all the DAACs, although not in a seamless or transparent manner. Instrument interdependencies, which require data products to be transferred between and distributed by several DAACs (e.g., Moderate Resolution Imaging Spectroradiometer [MODIS] products will be transferred among three DAACs), also draw the DAACs together.

The current consortium of DAACs includes eight discipline centers, although the SEDAC has never been fully integrated into the group. Each DAAC has a User Working Group whose membership is tailored to the mission and objectives of the DAAC. The DAACs, their host institutions, and their scientific specialties are listed in Table. 1.2.

Most of the DAACs were created from preexisting data operations. As a result of this heritage, each DAAC has its own information system for managing data and serving users, and holdings, sometimes going back decades. The DAACs' responsibilities within EOSDIS also vary, which has led to large differences in size, budget, and numbers of personnel among them. For example, the DAACs primarily responsible for managing data from the AM-1 platform and other near-term missions (GSFC, LaRC, and EDC) tend to have the largest staff

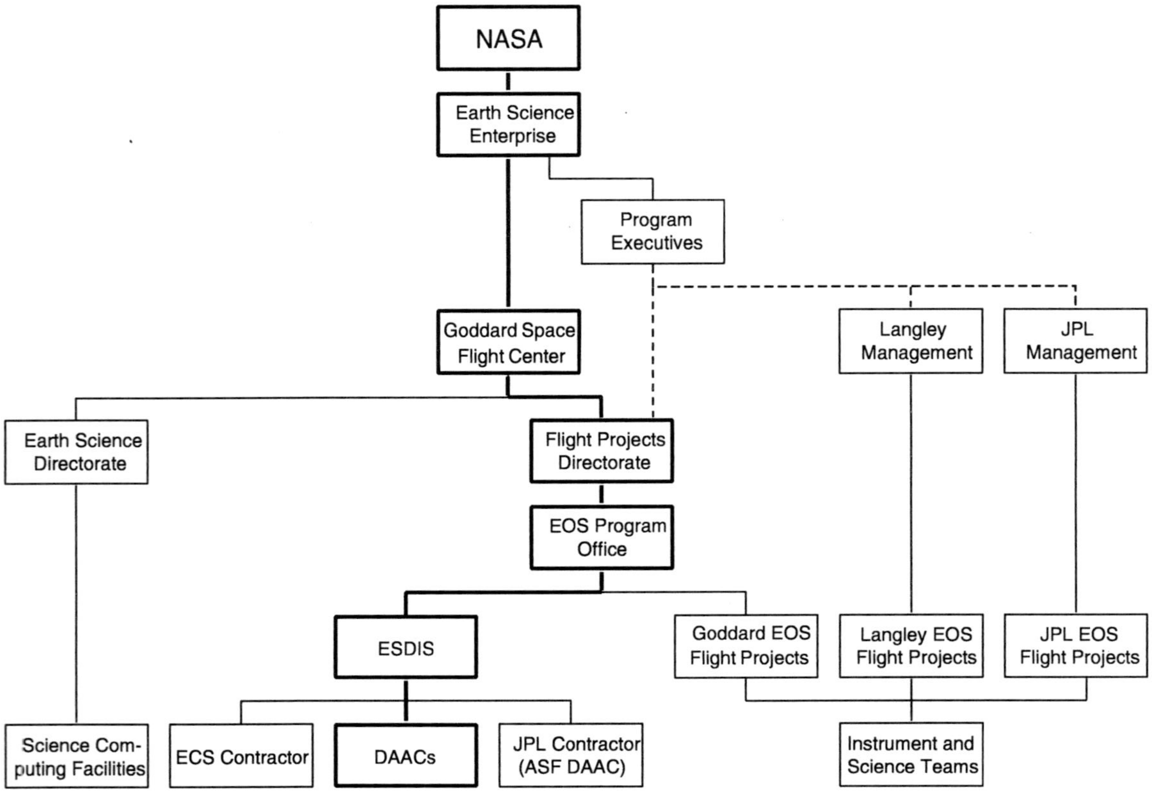

FIGURE 1.2. Reporting lines within EOSDIS as of December 1998. The program executives at NASA Headquaters are advocates for various programs at Goddard, Langley, and JPL, but do not provide oversight.

TABLE 1.2. NASA's Distributed Active Archive Centers

DAAC	Host Institution	Scientific Specialty
ASF DAAC	Alaska SAR Facility University of Alaska	Sea ice, polar processes
EDC DAAC	EROS Data Center U.S. Geological Survey	Land processes
GSFC DAAC	Goddard Space Flight Center NASA	Upper atmosphere, atmospheric dynamics, global biosphere, hydrologic processes
LaRC DAAC	Langley Research Center NASA	Radiation budget, aerosols, tropospheric chemistry
NSIDC DAAC	National Snow and Ice Data Center University of Colorado	Snow and ice, cryosphere
ORNL DAAC	Oak Ridge National Laboratory Department of Energy	Biogeochemical fluxes and processes
PO.DAAC	Jet Propulsion Laboratory NASA-Caltech	Ocean circulation, air-sea interaction
SEDAC	CIESIN Columbia University	Socioeconomic data and applications

and the most ECS-supplied hardware, software, and personnel (Table 1.3). DAACs that will receive data mainly from later missions (e.g., PO.DAAC and NSIDC) tend to have smaller numbers of personnel and lower budgets. (The NSIDC DAAC will not receive Level 0 data from the AM-1 platform, but will receive Level 2 AM-1 data pertinent to the cryosphere from the GSFC DAAC.) Table 1.4 lists the instruments, what they measure, and the anticipated launch dates of the EOS missions. The ORNL DAAC, which manages field-based observations rather than satellite measurements, is the smallest of the DAACs.

The ECS contractor (currently Raytheon Systems Corporation, formerly Hughes Information Technology Company) is responsible for designing an information system for DAAC functions ranging from ingest to archive. It has supplied the AM-1 DAACs (GSFC, LaRC, EDC, and NSIDC) with equipment, early versions of toolkits, software, subsystems, and personnel for installing and maintaining the ECS. The other DAACs are scheduled to receive more complete versions of the ECS software several years from now.

The DAACs played an advisory role while the ECS was being designed, but had no authority to change the functionality, implementation, or architecture of the system. Similarly, the DAACs have little control over how many ECS contractors are assigned to the DAAC to install the system or what tasks they will perform. Now that early versions of the software have been released, the DAACs

TABLE 1.3. DAACs at a Glance

DAAC	Number of Unique Users 1997[a]	Average Annual Budget ($M) FY 1994-2002[b]	Number of Staff FY 1998[c]	Current Holdings (TB)	AM-1 Instruments
ASF	400	DAAC: 6.0 JPL[d]: 6.6 ECS: 0.6	DAAC: 65 ECS: 1	110	None—foreign spacecraft: ERS-1,2; JERS-1; RADARSAT
EDC	1,156	DAAC: 5.6 ECS: 5.0	DAAC: 72 ECS: 38	9	Landsat 7, ASTER, MODIS
GSFC	12,216	DAAC: 5.1 ECS: 10.2	DAAC: 74 ECS: 40	4	TRMM, MODIS
LaRC	804	DAAC: 7.2 ECS: 3.6	DAAC: 84 ECS: 6	3	CERES/TRMM, CERES/AM-1, MISR, MOPITT
NSIDC	506	DAAC: 3.1 ECS: 1.1	DAAC: 27 ECS: 6	1	MODIS
ORNL	1,143	DAAC: 2.2 ECS: 0.1	DAAC: 14 ECS: 1	6-7 GB	None—data are mostly field based
PO.DAAC	15,527	DAAC: 4.6 ECS: 0.8	DAAC: 28 ECS: 1	15	None—US-foreign collaborative missions: AMSR, SeaWinds, Jason

NOTE: AMSR = Advanced Microwave Scanning Radiometer; ASTER = Advanced Spaceborne Thermal Emission and Reflection Radiometer; CERES = Clouds and the Earth's Radiant Energy System; ERS = European Remote Sensing Satellite; JERS = Japanese Earth Remote-Sensing Satellite; MISR = Multi-angle Imaging Spectroradiometer; MOPITT = Measurements of Pollution in the Troposphere; TRMM = Tropical Rainfall Measuring Mission.

[a]Includes only users who received data.

[b]DAAC, ECS, and JPL portions of the total budget are managed under separate contracts.

[c]DAAC includes DAAC staff and civil servants.

[d]Includes some non-DAAC data acquisition expenses.

are responsible for remaining informed about the contents of the releases, testing the system in an operational setting, and identifying bugs and missing capabilities. One or two ECS liaisons are assigned to each DAAC to facilitate the necessary two-way communication.

Instrument and science teams are responsible for developing algorithms for processing data from a particular instrument. The team either processes the data itself or transfers the algorithms to the DAAC for product generation. The DAAC then disseminates the data products and provides user services. This arrangement requires good communication between the instrument teams, science teams, and the DAACs, not only to ensure that high-quality products are produced, but also to ensure that the DAACs are sufficiently knowledgeable about the products to provide a high level of user services. Each EOS instrument—and, for that matter, each remote sensing instrument flown by NASA—is developed by an instrument team working in concert with at least one science team. Thus, each DAAC will have to interact with several teams, depending on the complexity of the instrument and the number of instruments collecting data at any given time.

The ESDIS Project, which is staffed by NASA engineers, computer programmers, and experts in budget and contract management, has several roles in the DAAC system. First, it is responsible for establishing the performance requirements to which the DAACs and the ECS contractor must adhere. The performance requirements were developed in consultation with the DAACs, the EOSDIS Panel, and the Mission to Planet Earth Program Office (now part of the Earth Science Enterprise). For the DAACs, the performance requirements focus on user service, data products, metadata, and processing. Second, ESDIS provides funding to the DAACs and evaluates requests for additional funds to create DAAC-unique extensions to the ECS. Such extensions permit the DAACs to support the specialized needs of their user communities. Similarly, requests for additional reprocessing or processing of nonstandard data products are evaluated by ESDIS. (The funding for the DAACs is managed separately from the ECS contract.) Third, ESDIS coordinates interaction between the DAACs and the ECS contractor. In the early stages of ECS development, questions or suggestions from the DAACs to the ECS contractor were routed through ESDIS. Now that the ECS is becoming operational, however, communications between the DAACs and the ECS contractor are more direct. Finally, ESDIS is responsible for the system-wide management and coordination of EOSDIS.

Alaska SAR Facility DAAC: A Unique Arrangement

The satellites contributing data to the ASF DAAC were launched several years ago. To manage the data, an information system had to be developed separately from the ECS, which was still in the design stages. The Jet Propulsion Laboratory was awarded the contract to develop processing, distribution, and archive capabilities for the ASF DAAC. (The JPL developers are separate from the PO.DAAC, which is also located at JPL.) In this sense, JPL plays the same role for the ASF DAAC as the ECS contractor plays for the other DAACs. JPL also serves as the instrument team for the ASF DAAC because it provides the processing algorithms.

Within ESDIS, a separate office has been established to oversee the ASF DAAC and the JPL developers. A more complete description of this relationship and its impact on DAAC operations is given in Chapter 6 (ASF DAAC).

TABLE 1.4. Future EOS and Related Missions

Satellite	Acronym	Measurement	Launch Date[a]
EOS AM-1 Platform			
• Clouds and the Earth's Radiation Energy System	CERES/AM-1	Measures the Earth's radiation budget and atmospheric radiation from the top of the atmosphere to the surface	March 1999
• Multi-angle Imaging Spectroradiometer	MISR	Determines planetary and surface albedo and aerosol and vegetation properties	March 1999
• Moderate Resolution Imaging Spectrometer	MODIS/AM-1	Measures biological and physical processes occurring on the surface of the Earth, in the oceans, and in the lower atmosphere	March 1999
• Advanced Spaceborne Thermal Emission and Reflection Radiometer	ASTER	Provides high-resolution images of the land surface, water, ice and clouds	March 1999
• Measurement of Pollution in the Troposphere	MOPITT	Measures thermal emissions to determine tropospheric CO profile and total columns of CO and CH_4	March 1999
EOS PM-1 Platform			
• Atmospheric Infrared Sensor	AIRS	Measures the Earth's atmosphere and surface to obtain temperature and moisture profiles	December 2000
• Advanced Microwave Sounding Unit	AMSU	Provides temperature and water vapor profiles for worldwide weather forecasting	December 2000
• Clouds and the Earth's Radiation Energy System	CERES/PM-1	Measures the Earth's radiation budget and atmospheric radiation from the top of the atmosphere to the surface	December 2000
• Moderate Resolution Imaging Spectrometer	MODIS/PM-1	Measures biological and physical processes occurring on the surface of the Earth, in the oceans, and in the lower atmosphere	December 2000
• Humidity Sounder, Brazil	HSB	Measures atmospheric humidity to improve weather prediction	December 2000
• Advanced Microwave Scanning Radiometer	AMSR	Observes atmospheric, land, oceanic, and cryospheric parameters, including precipitation, sea surface temperatures, and ice concentrations	December 2000

EOS Chemistry			
• Microwave Limb Sounder	MLS	Measures microwave thermal emission from the limb of the Earth's atmosphere to determine vertical profiles of atmospheric gases, temperature and pressure	December 2002
• Tropospheric Emission Spectrometer	TES	Measures the state of the Earth's troposphere and the interaction of ozone with other chemicals	December 2002
• High-Resolution Dynamics Limb Sounder	HIRDLS	Sounds the upper troposphere, stratosphere, and mesosphere to determine temperature, concentrations of oxides and aerosols, and the locations of polar stratospheric clouds and cloudtops	December 2002
Selected Complementary Missions			
• Tropical Rainfall Measuring Mission	TRMM	Monitors tropical rainfall and the associated release of energy that helps to power the global atmospheric circulation	November 1997
• Land Remote Sensing Satellite -7	Landsat-7	Provides the longest continuous record of the Earth's continental surfaces	February 1999
• Stratospheric Aerosol and Gas Experiment	SAGE III	Measures stratospheric aerosols, nitrogen dioxide, water vapor, and ozone	July 1999
• Active Cavity Radiometer Irradiance Monitor	ACRIM	Measures the Sun's total output of optical energy from ultraviolet to infrared wavelengths	October 1999
• Ocean Altimetry Mission—with France	Jason-1	Determines the circulation of the ocean and the distribution of sea surface topography	April 2000
• Earth Probe Total Ozone Mapping Spectrometer	TOMS	Maps the spatial distribution of total ozone, measures sulfur dioxide to detect volcanic eruptions, and measures the albedo of the Earth's atmosphere	August 2000
• SeaWinds—with Japan	SeaWinds	Measures wind vector fields near the sea surface	November 2000
• RADARSAT—Canada	RADARSAT-2	Measures topography and surface roughness, regardless of weather conditions.	November 2001
• Geoscience Laser Altimeter System	GLAS	Detects surface height changes in the ice sheets	July 2002
• Solar Stellar Irradiance Comparison Experiment	SOLSTICE	Provides long-term, accurate measurements of the solar ultraviolet (UV) and far ultraviolet (FUV) radiation	December 2002

[a]Schedule as of July 30, 1998.

SOURCES: Asrar and Greenstone (1995); http://www.earth.nasa.gov/missions/spacecraft.html.

EVOLUTION OF EOSDIS

A number of excellent histories of the EOS program and EOSDIS have been written (e.g., NRC, 1995b, 1998a). The following focuses on the history of EOSDIS, as it relates to the DAACs.

Original EOSDIS Concept

NASA's original plans called for the creation of a single DAAC to process, disseminate, and archive data from the entire EOS program, with the goal of creating "one-stop shopping" for researchers interested in studying the Earth as a system. However, NASA advisory groups, such as the EOSDIS Panel, objected that this arrangement would place too much responsibility in the hands of a single center, and recommended that data management functions be collocated with the relevant scientific expertise. A model for establishing geographically distributed discipline centers was first described in a 1986 report of the National Research Council's Committee on Data Management and Computation. The committee envisioned a set of active database sites, which would receive regular scientific use and guidance by associated scientists in the corresponding discipline (NRC, 1986). Such sites differ from data centers in that they exist for a fixed period of time—the period when the data are being used intensively for research—and are thus not responsible for long-term maintenance of the data (see Chapter 2, "*DAAC Versus Data Center*"). NASA adopted the model, and by the early 1990s, eight DAACs had been established by program leaders at NASA Headquarters, and the ninth, SEDAC, was established by congressional fiat shortly afterward. (The DAAC at Marshall Space Flight Center was closed due to budgetary pressures in 1997.)

In a multi-DAAC system, one-stop shopping meant that users would be able to access the system through any DAAC, search all the EOSDIS holdings, and obtain the relevant data, regardless of where it resided. The use of common formats (e.g., Hierarchical Data Format [HDF]-EOS) and standards across the DAAC system would permit users to integrate data of different types with a wide range of temporal and spatial scales. To test these concepts, the DAACs participated in two key prototype exercises—the Pathfinder Program and EOSDIS Version 0. Pathfinder data sets were developed by science teams to support global change research and to gain experience in reprocessing and transferring massive data sets in the pre-EOS era. Because Pathfinder products incorporate data from many disciplines (land, ocean, and atmosphere), sources (NASA, the National Oceanic and Atmospheric Administration [NOAA], U.S. Geological Survey [USGS], and Environmental Protection Agency [EPA]), and spatial and temporal scales, the program also illustrated some of the difficulties inherent in integrating disparate data types.

Version 0 was developed largely from existing hardware and software at the DAACs. It was designed to provide an early operational capability and to test selected EOSDIS tools and services. However, Version 0 had limited ingest, processing, and archive capabilities. These capabilities were to be provided in a comprehensive new information system, the ECS, which would replace Version 0. A contract to develop the ECS was awarded to a single contractor (Hughes Information Technology Company) in 1993.

Early peer reviews found that the system being developed by the ECS contractor would likely be too rigid to permit users to manipulate the data in new ways or to evolve to meet new user needs (e.g., NRC, 1994). A 1994 NRC report concluded that the DAACs were in a good position to understand the needs of their users and should therefore become intimately involved in the development of the ECS. Their involvement would help ensure that the information system supported the scientific community for which it was built (NRC, 1994). This recommendation was never implemented.

Federation and Recertification

Although ESDIS and the ECS contractor took steps to make the information system less centralized and more flexible, rapid technological changes called into question the original EOSDIS paradigm. In particular, the growth of the World Wide Web (WWW) and the widespread availability of powerful desktop computers made it possible for individuals to manipulate and store large data sets for the first time. (The scale of the data management problem is discussed in Box 1.1.) As a result, many traditional data management tasks no longer have to be performed by DAACs or data centers, and a more truly distributed system for EOSDIS could be created. With this goal in mind, an NRC committee recommended that certain DAAC functions, such as product generation, publication, and user services, be transferred to a federation of partners selected competitively from academia, industry, and government (NRC, 1995a). Given the imminence of the AM-1 launch, the committee also recommended that NASA federate EOSDIS in stages, beginning with an initial limited set of pilot projects (NRC, 1996). In 1998, NASA initiated a prototype federation with participation by three types of Earth Science Information Partners (ESIPs). The Type 2 and Type 3 ESIPs (see Box 1.2) were selected through a competitive process to create products and offer services not currently provided by EOSDIS (see NRC, 1998b); the DAACs represent Type 1 ESIPs. The responsibilities of the ESIPs are described in Box 1.2.

The issues likely to be faced by the ESIPs in creating the prototype federation were examined at a 1998 NRC workshop. The report from that workshop examined the federation concept; compared governance models from a diversity of federated structures—libraries, international organizations, industry, and academia; and offered some lessons for managing scientific data in an ESE federation (NRC, 1998b). A new NRC report about to be released (NRC, 1998a)

BOX 1.1. Scale of the EOSDIS Data Management Problem

This report uses a number of terms relating to the size of EOSDIS data sets. These terms are relative and reflect the committee's view of the manageability of data at the time of writing.

A major concern in early EOSDIS planning was the sheer size of the data streams that must be processed routinely and the availability of adequate computer power and communications bandwidth to handle them. As the technology and installed infrastructure have improved, the emphasis has shifted to distributed operations and the interface to scientific users, in particular for the science teams overseeing the development and production of standard products. However, it remains true that the data volumes from the AM platform will be unprecedented, and approaches that are normal for a desktop workstation may not be applicable on the scale of EOSDIS operations. What is "routine" is very much a shifting target, but discussions of strategy must reflect realistically the orders of magnitude involved. Based on the survey of users (see Appendix D), a typical user request for data from a DAAC is as follows:

- small: <10 Mbyte,
- typical: 10 Mbyte to 1 Gbyte, or
- large: 1-100 Gbyte.

These distribution limits are determined primarily by ease of transmission over the Internet and by standard capabilities on workstations and personal computers. One gigabyte of data can be fitted onto two CD-ROMs. Nevertheless, any assessment of such scaling issues should be cognizant of the fact that current Internet bandwidth is doubling every three months or so.

From the perspective of a scientific data center, data sets could be characterized as:

- small: 1 Tbyte,
- large: 100 Tbyte, or
- very large: 1 Petabyte.

These limits are determined primarily by the availability of mass storage systems and associated data management software. One Petabyte would require a stack of CD-ROMs several kilometers high! Of course, the effort required to handle a data set effectively depends on many factors in addition to its size, such as the complexity of its structure, patterns of use, and user understanding of its content.

BOX 1.2. Responsibilities of the Prototype Federation Partners

Type 1 ESIPs. These ESIPs are responsible for standard data and information products whose production, publishing or distribution, and associated user services require emphasis on reliability and adherence to schedules. Type 1 ESIPs include DAACs and science teams for specific instruments.

Type 2 ESIPs. These ESIPs are responsible for producing innovative science information products and services, which primarily serve the global change and earth science communities. Type 2 ESIPs include science teams and global change scientists.

Type 3 ESIPs. These ESIPs are responsible for providing innovative, practical applications of earth science data to a broad range of users beyond the global change research community. Type 3 ESIPs include science teachers, college earth science students, policy analysts, interested public, research scientists working outside their discipline, and for-profit businesses.

SOURCE: Modified from NRC (1998b).

will provide additional recommendations on refining the ESE federation model and the responsibilities of the partners.

If the prototype is successful, NASA plans to phase in development of an Earth Science Enterprise federation through a series of competitions focusing on production, publication, and user services, following launch of the AM-1 platform and other near-term missions. Additional DAAC-type activities may also be competed. Only recertified DAACs will be permitted to compete for these functions and, if successful, become partners in the ESE federation. Consequently, a DAAC recertification process may accompany each competition.

Recent Developments

As the previous discussion illustrates, plans for implementing EOSDIS have undergone considerable change over the past five years. Further change, particularly in the EOSDIS Core System, is likely to occur as the system approaches operational readiness. Delays in the ECS led NASA's 1997 Biennial Review Committee to recommend stronger managerial oversight of the ECS, the creation of backup plans, and a reduction in the data requirements for EOSDIS (Independent External Review Panel, 1997). A subsequent demonstration of the ECS showed that full functionality of the system would likely not be achieved in time for launch of the first EOS satellite, the Tropical Rainfall Measuring Mission (TRMM). Consequently, NASA decided that the DAACs would still plan to generate Level 1 products beginning 90 days after launch, but that production of higher-level products would be delayed. Only 25% of the standard Level 2, 3, and 4 products would be produced the first year, reaching 100% in the fourth year after launch. This strategy became known as the 25-50-75 scenario. NASA also solicited backup plans from the DAACs and science teams for each instrument. The backup systems of the GSFC and LaRC DAACs are already being used to manage data from TRMM, which was launched in November 1997.

Meanwhile, NASA directed the ECS contractor to focus system development on preparing for the AM-1 data streams. A failure in the flight operations segment of the ECS (under subcontract to a different developer) in April 1998 delayed launch of the AM-1 platform and gave the ECS contractor at least six more months to prepare. By July 1998, however, it became clear that these efforts would not be sufficient, and NASA is now considering using backup plans more extensively. The backup systems developed by the DAACs rely on Version 0 and their own home-grown information systems to process the data. On the other hand, backup plans developed by the science and instrument teams mostly call for the data to be processed at the Science Computing Facilities, where the algorithms were developed and where the interdisciplinary science teams will eventually analyze the data. Which plans are implemented will be decided on an instrument-by-instrument basis, but in either case, the ECS would be used only for data distribution and archive.

ORGANIZATION OF REPORT

In this report the Committee on Geophysical and Environmental Data examines the DAAC system as it was configured from October 1997 to September 1998, and those elements of EOSDIS that pertain directly to the DAACs. Its assessment, presented in Chapter 2, draws on the reports of the DAAC review panels, interviews with system developers and NASA management, and a survey of DAAC users.

The individual DAAC reports were written by specialized site visit panels and are presented without modification by the CGED. They are given as Chapters 3 through 9, and they appear in the order in which the DAACs were visited. Each report is based on a site visit, subsequent e-mail discussions with DAAC personnel, and the personal experiences of panel members with the center. The site visit agenda followed by all the panels is given in Appendix A. For consistency the DAAC reports follow a similar format. The chapter sections reflect the criteria for review, which are divided into five categories: holdings, users, technology, management, and the relationship between the DAAC and other components of the Earth Science Enterprise. The criteria for review and suggested measures of performance appear in Appendix B. In each chapter, an abstract provides an overall assessment of the DAAC and identifies the panel's key recommendation(s). Throughout the report, suggestions are made for improving the effectiveness of the DAACs, but only the most important are phrased as recommendations to NASA, ESDIS, or the DAAC.

To prepare the panels for their discussion of the relationship between the DAACs and the Earth Science Enterprise, the CGED interviewed ESDIS in advance of the site visits. Questions prepared by the CGED and the formal written responses of ESDIS are given in Appendix C. The questions were based on the strategic management plan for the DAAC system, which is prepared yearly by the DAACs and ESDIS, and focus on the following topics: (1) ESDIS expectations of the DAACs and (2) DAAC expectations of ESDIS. In addition, Appendix C includes the ESDIS response to issues raised by informal advance teams who visited the DAACs and by previous NRC reports.

The committee solicited input from the broader community by conducting a user survey. The survey was e-mailed to the Investigator Working Group list, which includes nearly 1,000 individuals associated with EOSDIS and the Earth Science Enterprise, including instrument and science teams, program managers, and NASA advisory panels. It was subsequently forwarded to other users in the United States and abroad. The results of the survey can be found in Appendix D. Nearly 400 users responded, including scientists, educators, and the general public. The survey was not rigorously controlled; therefore the committee did not perform a statistical analysis of the results. Nevertheless, the patterns of responses shown in Appendix D illustrate the range of experiences that different users groups have had with the DAACs.

Finally, an acronym list, which defines the many organizations, satellite missions, and science projects discussed in this report, appears at the end of the report.

2

Overview of the DAAC System

This chapter synthesizes overarching issues that were noted, with varying perspectives, in the individual DAAC reviews. The issues are related to (1) the DAACs as a system, (2) the need for flexible approaches for implementing EOSDIS, (3) the relationship between the DAACs and their users, (4) life-cycle data management, and (5) the role of NASA.

WHAT IS A DAAC? WHY A SYSTEM?

DAAC Versus Data Center

As noted in Chapter 1, the DAACs were created to be different from data centers. Data centers are permanent facilities—their primary focus is on long-term distribution, maintenance, and archive of data and data products. On the other hand, DAACs are meant to exist for only about 15 years and to be involved in the initial, active stages of a satellite program, when the most intense scientific activity is occurring. (The DAACs also have heritage data sets, sometimes going back decades, from preexisting data operations.) Key DAAC tasks include (1) supporting the operational ingest and management of a suite of spaceborne sensors operated as part of the Earth Science Enterprise, (2) producing data products from remotely sensed and complementary *in situ* data sets as required, and (3) reprocessing data in response to improvements in the algorithms or to correct errors detected in the processing. Providing user services and access to the data is important to both DAACs and data centers. Consequently, the DAACs must operate according to sound principles of data center management (see "*Life-*

Cycle Data Management," below). To fulfill their DAAC mission, however, they must also be responsive to scientific needs in a dynamic environment.

The ORNL DAAC neither manages satellite data nor works with EOS science and instrument teams, even though its *in situ* data sets are critical for calibration or validation purposes. Thus, by the definition given above, the ORNL DAAC operates more like a data center than a DAAC. Similarly, if the backup plans of the science teams are adopted, few DAACs will ingest or process EOS data. Those that do not will no longer be DAACs in the sense originally envisioned by NASA.

The DAAC System

The DAACs have a dual role within the Earth Science Enterprise. Not only do they operate as discipline centers that serve the needs of a relatively small, specialized constituency, they cooperate as elements of a larger system, which serves the broader earth science community. The former must be a primary role—otherwise the DAACs cannot operate as effective data centers. The latter is an additional responsibility of the EOSDIS DAACs. Fulfilling the second of these roles is difficult because the DAACs are profoundly different from one another. Comparison of Chapters 3 through 9 indicates the following differences among the DAACs:

- **Their core constituencies are different.** The discipline focus is different for each DAAC (Table 1.2), but even DAACs with overlapping disciplinary interests serve a distinctive set of users. For example, the ASF DAAC serves sea-ice scientists interested in synthetic aperture radar data, whereas the NSIDC DAAC serves the broader polar science community. Similarly, the EDC and ORNL DAACs serve terrestrial ecologists, but the EDC DAAC focuses on users of remote sensing imagery, and the ORNL DAAC focuses on users of *in situ* data from field campaigns and process studies.
- **Different disciplines place different demands on the information system.** For example, the cryospheric studies facilitated by the NSIDC DAAC require polar projections, and the field-based data of the ORNL DAAC require a broader metadata model than would be developed for remote sensing data alone. In addition, the standard EOSDIS data format, HDF-EOS, is poorly suited for ASF and ORNL DAAC holdings, and the three HDF data structures supported by the ECS—point, swath, and grid—do not apply to all LaRC DAAC data.
- **They are hosted by a diverse array of institutions**. The GSFC, LaRC, and EDC DAACs are housed in government-operated facilities, the ORNL DAAC and the PO.DAAC are located in facilities managed by private institutions, and the ASF and NSIDC DAACs are housed in universities (Table 1.2).
- **They vary in size.** In terms of the size of the budget and the number of staff, GSFC is among the largest and ORNL is the smallest of the DAACs (Table

1.3). The ASF DAAC has the largest volume of holdings, and the ORNL DAAC has the largest number of data sets. The GSFC DAAC and the PO.DAAC have the largest user base.

- **They have different readiness requirements.** The GSFC, LaRC, and EDC DAACs must focus on preparing for the massive data flows from near-term missions such as the AM-1 platform. On the other hand, the NSIDC DAAC and the PO.DAAC will primarily manage data from later missions and can continue to focus on refining their service to existing users. Preparation time for the ASF DAAC, which has been receiving large volumes of data for several years, has already passed.
- **Each has developed unique systems for managing data.** For example, the GSFC DAAC developed the Archer file management system. In addition, the GSFC and LaRC DAACs developed information systems to handle TRMM data (TRMM support system and the Langley TRMM Information System, respectively). Comparable variability exists in the computing, storage, and communications hardware.
- **Manipulating the data requires varying levels of technological sophistication from users.** GSFC, ASF, and EDC DAAC users typically deal with very large data sets and require substantial computer resources to work with and analyze the data. They also require high-capacity media distribution. In contrast, ORNL DAAC users deal with many small ASCII tables, which are easily transmitted over ordinary Web channels and manipulated using standard personal computer software.

In addition, the ASF DAAC differs from the other DAACs in three important ways. First, it is managed by ESDIS separately from the other DAACs. Second, its processing, distribution, and archive systems are being developed under a contract to the Jet Propulsion Laboratory rather than by the ECS contractor. This is a consequence of the ASF DAAC becoming operational far in advance of the other DAACs. Third, it manages data collected exclusively from foreign spacecraft. The space agencies that operate the satellites have placed severe restrictions on the amount and geographic coverage of data that U.S. researchers can obtain at affordable prices. All other EOSDIS data are available at no more than the cost of filling a user request.

EOSDIS, however, is meant to be more than a collection of discipline centers. To fulfill their mission to serve the broader earth system science community, the DAACs must adopt a mind-set that they are also components of a coherent (but not necessarily uniform) system that (1) enables users to locate, access, and use various types of data with valuable scientific content, using a common set of tools, whatever the data type; and (2) stimulates collaborative, multidisciplinary research as a means for understanding earth system science processes. Although issues in interpreting and blending multisensor time-varying data sets at different resolutions and sampling strategies are unavoidable, comparable access technologies, a consistent terminology, and unobtrusive authorization procedures are key aspects of an information environment that fosters rather than hinders such integrative inquiry. At present, this outcome could be achieved by a coherent system of DAACs, although this management model may be superseded in the future by a federation of partners (see "*The Need for Adaptability,*" below).

By working together, the DAACs can take advantage of tools and technologies developed at other centers, rather than creating everything in-house. They also enjoy the benefits of collective bargaining with NASA management. Finally, as recent history has shown in the case of highly visible events (e.g., the Mars Pathfinder mission), the Web environment is particularly prone to extraordinary surges in user demand for certain types of data. This is another reason why the DAACs might profitably form an alliance to foster growth of their collective user base and to ensure a reliable level of service through the crises that are sure to arise.

None of the panels detected significant coordination among the DAACs. A noteworthy exception is the User Services Working Group, which identifies and analyzes problems collectively and devises innovative solutions on a system-wide basis. The experience gained in this process, as well as past experience in developing Version 0 of EOSDIS, should provide a practical basis for creating a working system. However, the profound differences between the DAACs have made their integration into a seamless distributed system a daunting task, whose completion has largely eluded them so far. Moreover, the need to develop DAAC-specific systems for handling the AM-1 data streams is driving the DAACs further apart (see "*The Need for Adaptability,*" below). Finally, a less tangible, but perhaps more difficult, barrier to overcome is the apparent reluctance of many of the DAACs to become a part of EOSDIS. It is clear from the panel reports that the ECS contractor failed to take the DAACs' views into account when designing the system. Thus, the DAACs do not have a sense of ownership in the ECS. Moreover, efforts to coordinate DAAC activities in the past seem to have been motivated more by strong leadership at ESDIS and NASA Headquarters than by self-interest or a belief in the goals of the system. Many of the DAACs feel that the long-range vision of Dixon Butler, former operations director of the Data and Information Systems Division of Mission to Planet Earth, and Gregory Hunolt, former DAAC system manager, is not shared by current managers. For the system to work, there needs to be a reaffirmation of EOSDIS goals at all levels—NASA management, the DAACs, and the scientific community.

If a coherent system cannot be achieved, users will have to learn to deal with a disparate system of DAACs. In doing so, users may benefit from the results of NASA's prototype federation of Earth Science Information Partners (ESIPs). Indeed, some of the ESIPs were funded to find ways to link disparate data repositories and management centers together, and they are reportedly making progress. Through requests for proposals, the federation mechanism may promote innovative and effective ways to help users access data from a wide range of sources (including the DAACs) easily and efficiently.

Redundancy in the DAAC System

It is probably intrinsic to the nature of distributed systems that they should suffer from some degree of redundancy. However, reckless tracking and elimination of redundancies may lead to a monolithic architecture that rapidly loses the ability to evolve. It is clear from the following chapters that each DAAC occupies an important, unique niche in the Earth Science Enterprise. None of the panels raise concerns about functional overlaps between DAACs or about other possible redundancies. In fact, one is hard pressed to find a clear-cut case in which a DAAC's function could easily be eliminated without incurring a worrisome loss to the science. Even the ASF and ORNL DAAC panels, which suggest fundamental changes to DAAC operations, argue strenuously for the scientific worthiness of these DAACs. The fact that the DAACs are housed within science facilities, where the appropriate scientific and data management expertise is readily available, is a strength of EOSDIS. Mergers between DAACs can be devised that might lead to economies of scale, but they would also lead to a loss in expertise and talent. The cost of this loss in expertise, although difficult to quantify, should not be underestimated.

This argument holds true particularly in the case of DAACs that have achieved a high level of symbiosis with their user communities. The PO.DAAC is a case in point. JPL's efforts to outsource the operation, as a money-saving measure, would destroy the highly productive team spirit that has been carefully nurtured between the DAAC and its collocated science users. The PO.DAAC panel's central recommendation, to leave the DAAC embedded within the physical oceanography group at JPL, is not addressed to the DAAC, but to JPL and ESDIS.

Conclusions and Recommendations

The committee concludes that there is no obvious redundancy in the roles and responsibilities of the DAACs and that each DAAC has a critical role to play in the overall endeavor. However, the DAACs do not as yet function as a system, although they coordinate operations on a limited scale. The creation of a coherent, seamless system in the sense that was envisaged originally will require (1) a reaffirmation of the scientific advantages of functioning as a system, (2) a collective effort to counter the pressures driving the DAACs apart, and (3) more serious efforts to collaborate. Otherwise, there is real danger of losing coherence, no matter how many teleconferences and meetings are held by the managers. If a coherent system of DAACs cannot be achieved, it is incumbent on NASA to provide the necessary resources (e.g., through requests for proposals to participate in NASA's federation) so that a set of disparate data centers can serve the multidisciplinary users as effectively as an ideal DAAC system.

Recommendation 1. The DAACs do not yet act as components of a coherent system. They share the responsibility for providing the vision and leadership toward this goal with the science teams, ESDIS, and NASA. If such a coherent system cannot be achieved, NASA should place a greater emphasis on user-generated proposals seeking to help the community deal with a disparate DAAC system.

Recommendation 2. A DAAC alliance with a common goal will better serve the broader community than the collection of individual centers that currently exists. The DAACs should support each other and express a collective point of view on EOSDIS policies.

THE NEED FOR ADAPTABILITY

A Changing Paradigm

EOSDIS is evolving rapidly. The changes are being driven by (1) delays in the ECS, (2) a rapidly changing network environment, and (3) new management approaches to EOSDIS. In the current paradigm that governs the DAAC system, the ECS is supposed to provide the glue for the system—the layer of uniformity that presents a seamless appearance to the users. Uncertainties concerning the ECS—in terms of both performance and delivery schedule—have led to irresistible pressure to turn to local, DAAC-specific solutions, which are not always transparent to the users. Such solutions are not necessarily bad—the development of DAAC-specific information systems was necessary for managing data from the TRMM mission—but they do pose a challenge to creating a coherent system.

A question therefore arises: Is EOSDIS a concept that has been overcome by events, so that its time has passed?

The committee believes that technological advances and new management approaches offer hope for achieving the ultimate goal of an integrated DAAC system, albeit by taking a completely different route than originally envisaged by its architects. For example, the Web provides a new way to link the holdings of the centers and to make data easy to find and access. All of the DAACs are well on their way to developing Web-based interfaces for their users, although these efforts still fall short of the full complement of capabilities that the panels deem desirable. Improvements are needed in the realms of tracking users, data requests, and center performance.

Under the federated EOSDIS paradigm, the diversity of the DAACs is a good thing because it allows them to satisfy the needs of equally diverse groups of users. For the federation to work, however, the newly found flexibility must be embraced by the system architects. The panels advocated different approaches to the system architecture of the DAACs:

- the GSFC DAAC should continue working with the ECS;
- the LaRC DAAC should work to make its systems compatible with the ECS; and
- the PO.DAAC should adopt only the ECS components needed to make the system work.

The committee concurs with the concept of multiple architecture strategies, noting however that their implementation would make a seamless system more difficult to achieve. Nevertheless, by taking advantage of Web technologies and making a concerted effort to act as components of a system, it should be possible for the DAACs to realize this goal.

Conclusions and Recommendations

The CGED agrees with previous NRC committees (e.g., NRC, 1998a) that the new federated paradigm should meet the stated EOSDIS goals, provided the architecture is implemented in a flexible way. Indeed, with the adoption of DAAC-specific approaches to managing the AM-1 data streams, the increased participation of EOS science and instrument teams in processing data, and the initiation of the prototype federation, the move to an EOSDIS federation is timely and probably inevitable. Instead of requiring the DAACs to accept and implement the ECS as a complete system, the committee believes they should be permitted to select those subsets of the ECS that allow them to function most effectively as components of an adaptable system and to flow with the rapidly changing electronic data environment. With this flexibility, however, comes a clear responsibility: the DAACs will have to earn their title and their place in the system by discharging their duties as system components. Otherwise, they risk turning into relatively trivial custodians of the data sets in their charge or being replaced.

Recommendation 3. To take advantage of the unprecedented flexibility afforded by the new Web-based technologies, ESDIS should allow the DAACs to incorporate only those components of the ECS that they require to satisfy their user community. This flexibility should not come at the cost of reducing the DAACs' ability to function as full-fledged members of the DAAC system. For the DAACs, the price of this flexibility is an increased individual responsibility to contribute to the overall goals of EOSDIS.

DAACS AND THEIR USERS

The DAAC system is the principal element of EOSDIS through which the ESE interacts with its various constituencies. If the users are unable to obtain the data they need in useful form and in a timely manner, the ESE will fail. Therefore, the various panels and the committee devoted considerable attention to DAAC users.

User Community

As described in the recent NRC study on an Earth Science Enterprise federation (NRC, 1998b) the principal ESE constituencies include:

- data producers, including instrument teams and scientists conducting *in situ* studies (e.g., scientists contributing data to the ORNL DAAC);
- global change scientists, who use and synthesize a broad range of data from different sources, and who may also produce higher-level data products;
- knowledge brokers, including policy makers, teachers, students, and the interested public, who use reliable, interpreted data products or assessments; and
- for-profit businesses, which generate value-added data products for commercial purposes.

According to the guidelines adopted by NASA Headquarters, EOSDIS will support the following constituencies, which overlap with the ESE constituencies listed above:

- national and international agencies and entities with whom NASA has written agreements or legal obligations concerning ESE data;
- NASA-funded ESE investigators (e.g., data producers, global change scientists);
- the broader U.S. and international earth science community (primarily global change scientists); and
- U.S. policy makers (i.e., knowledge brokers).

If sufficient resources are available, EOSDIS will also support other knowledge brokers, including the U.S. education community, the U.S. general public, and other interested users (see Appendix C). Given that the Earth Science Enterprise is a science program, the committee agrees with these broad priorities, noting that most DAACs have a substantial outreach activity (see below).

Most panels report that the DAACs assign their highest priority to data producers, global change scientists, and NASA's partner agencies. The data producers are important to the DAACs because they generate a significant fraction of the DAAC holdings. The science community, often labeled the "customers," is considered the primary user group. Failure to satisfy this user group would constitute a failure of the DAACs (although one must be careful not to attribute all shortcomings of EOSDIS to the DAACs). However, the committee feels that the best data centers (and DAACs) go beyond the minimum requirements and that failure to serve a broader community should be considered less than stellar performance. Some DAACs anticipate that for-profit businesses, although not a high-priority NASA constituency, will be a fast-growing segment of the user community. For example, the NSIDC DAAC sees a growing constituency among geotechnical engineers in permafrost areas, notably in other countries. Sea-ice products produced by the ASF DAAC are important to the shipping industry. Finally, potential users of the EDC DAAC are far more likely to be interested in commercial applications of Landsat than in scientific research. To reach the knowledge brokers, the DAACs sponsor outreach activities. Outreach is typically targeted at K-12 educators but also includes dissemination of information to the general public via the Web and a variety of media (e.g., brochures, flyers), as well as displays at conferences. In addition, it includes the dissemination of near-real-time data of general interest (whenever appropriate) via the Web and the production of data sets in a variety of popular formats.

Although most DAACs are aware of the broad characteristics of their user communities (e.g., U.S. versus foreign user, scientists versus K-12 educators), few have a detailed understanding of their user profiles. Without a detailed understanding of who their users are and how they use the data, the DAAC will not be able to provide the specialized services necessary to get the most out of the data. Moreover, it will be more difficult for the DAACs to expand their user base. Consequently, most of the panels advise their respective DAACs to characterize their user community more quantitatively, to track its evolution, and to incorporate this information in performance metrics and in a strategic plan. It is important that such tools be implemented as early as possible, before the massive influx of EOS data.

The problem of characterizing the user community is compounded by the fast-growing use of the Web. Tracking users who access the DAAC as casual browsers through the Internet is notoriously difficult, but this problem is faced by all providers of products and services on the Web, including commercial providers, so innovative solutions are bound to emerge. Some of these solutions will likely be applicable to the DAACs. Instituting log-in procedures is certainly possible, but such procedures place a high administrative burden on the DAAC and discourage casual browsers. Consequently, several DAACs have generally decided against log-in procedures, except for restricted data sets generated by foreign sources. In addition, national data centers have developed strategies for keeping track of the user communities they serve, updating user profiles, and soliciting user input on the usefulness of their products. Their experience would likely provide a useful guide to the DAACs.

User Survey

Although not the only metric by which to measure the performance of a data center, user satisfaction is central to its long-term health, and dissatisfaction among a measurable fraction of the user base is a sure sign of a dysfunctional center.

Several DAACs have conducted user surveys in recent years. Some are superbly comprehensive and informative, such as the most recent ASF DAAC survey mentioned in Chapter 6. Nevertheless, these surveys do not address the systemic aspects of the DAACs. To reduce this gap, the committee conducted its own informal survey, focused on the most important category of users, namely the scientific community. No effort was made to achieve a statistically reliable sampling. Rather, an electronic questionnaire was mailed to the entire membership (~1,000) of the EOS Investigators Working Group (IWG), and survey recipients were free to forward the questionnaire to other users. For instance, the PO.DAAC secured additional responses from a large contingent of users outside the IWG, including many educators and users from abroad. Consequently, the answers should be taken only as a qualitative assessment of opinion trends among users.

The survey and responses (393 responses, including 184 from foreign users) are provided in Appendix D. The responses were divided into three categories: (1) sophisticated users, including members of the IWG and data providers; (2) casual users, including U.S. educators and individuals who used a DAAC a few times a year or less; and (3) foreign users, including scientists and graduate students. Basic patterns of interest are the following:

- Only about one-third of sophisticated users obtain data from a single DAAC. Most use several DAACs, sometimes as many as five. Casual and foreign users, on the other hand, tend to use a single DAAC.
- As illustrated in Figure 2.1, the majority of survey respondents declare themselves to be satisfied with the DAACs, and judge their performance to be above average. The only significant dissatisfaction was expressed—quite emphatically—by scientific users of the ASF DAAC and, to a lesser extent, the EDC DAAC. The major sources of strong dissatisfaction, described in sometimes long essays, had to do with (1) slow response; (2) difficulties in finding data; and (3) poor user services, particularly in tracking data requests. Significant problems with user services at these DAACs were also noted by the corresponding review panels. It is noteworthy that none of the casual or foreign users expressed anything but satisfaction with the system, sometimes in rather eloquent and glowing terms. For these users, the fact that the data are free and unrestricted far outweighs any difficulties they may have in obtaining them.
- A majority of survey respondents claim to access the DAACs several times a year. A sizable minority access the DAACs on a weekly or even daily basis.

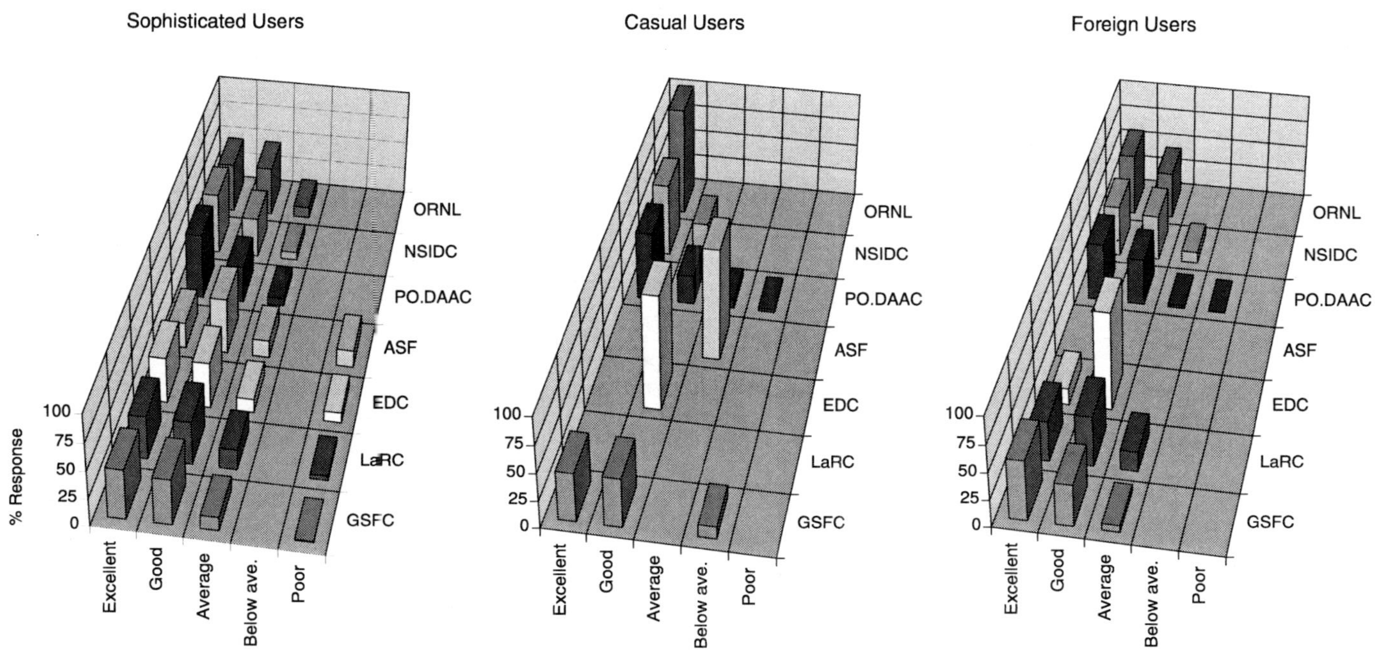

FIGURE 2.1. User satisfaction. SOURCE: user survey (see Appendix D).

• The volume of data requested is typically 10 MB to 1 GB. Casual and foreign users tend to user smaller data sets than sophisticated users. Almost all users claim to acquire data both through electronic transfer and via distribution media.

• Ease of access is judged to be above average, with a majority of survey respondents considering access to be "somewhat easy" to "very easy" (Figure 2.2). Foreign users had more difficulty accessing the DAAC's holdings than sophisticated and casual users. Again, however, a vocal minority of scientific users reported that access to the ASF and EDC DAACs is "very difficult."

As noted in the second bullet, a minority of survey respondents have difficulty finding data in usable forms. Surprisingly, many of these users are scientists who are familiar with the EOS missions and instruments and with the data sets managed by the DAACs. The time spent by potential users in figuring out how to use the DAAC system is potentially large from the users' point of view, and this cost is not accounted for. For these users, the system has a high cost of entry. The committee feels that each DAAC should analyze the cost of entry to users with the goal of bringing it down significantly. The GSFC DAAC has done this in its business model, which strives to decrease the effort of "customers" by offering data sets in manageable and digestible chunks.

The Role of Scientists

Previous data center reviews suggest that the most effective data centers incorporate active scientists into their day-to-day operations, either (1) as consultants for dealing with user queries; (2) as friendly users who will "tire-kick" new data sets or tools on behalf of the center; or (3) as advisers on center priorities and opportunities for initiatives. In addition, some data centers have active full- or part-time scientists on staff; having first-hand experience and being able to contribute to the overall quality control process of data sets can be scientifically rewarding. Such on-site feedback is an essential part of building and sustaining an information system for science. It is at least as important for achieving the goals of the center as the technical aspects of computer software and hardware.

The day-to-day involvement of scientists is even more important to the DAACs because they are involved in active scientific investigations and therefore must be more responsive to scientific needs than typical data centers. To help them do so, the DAACs are all located within facilities where relevant scientific experience is close at hand, and each has a User Working Group comprising external scientists to advise it on priorities and scientific needs. The DAACs take advantage of this expertise in varying degrees. For example, DAACs with a good track record of implementing the suggestions and recommendations of their User Working Groups include the PO.DAAC and the EDC, NSIDC, and ORNL

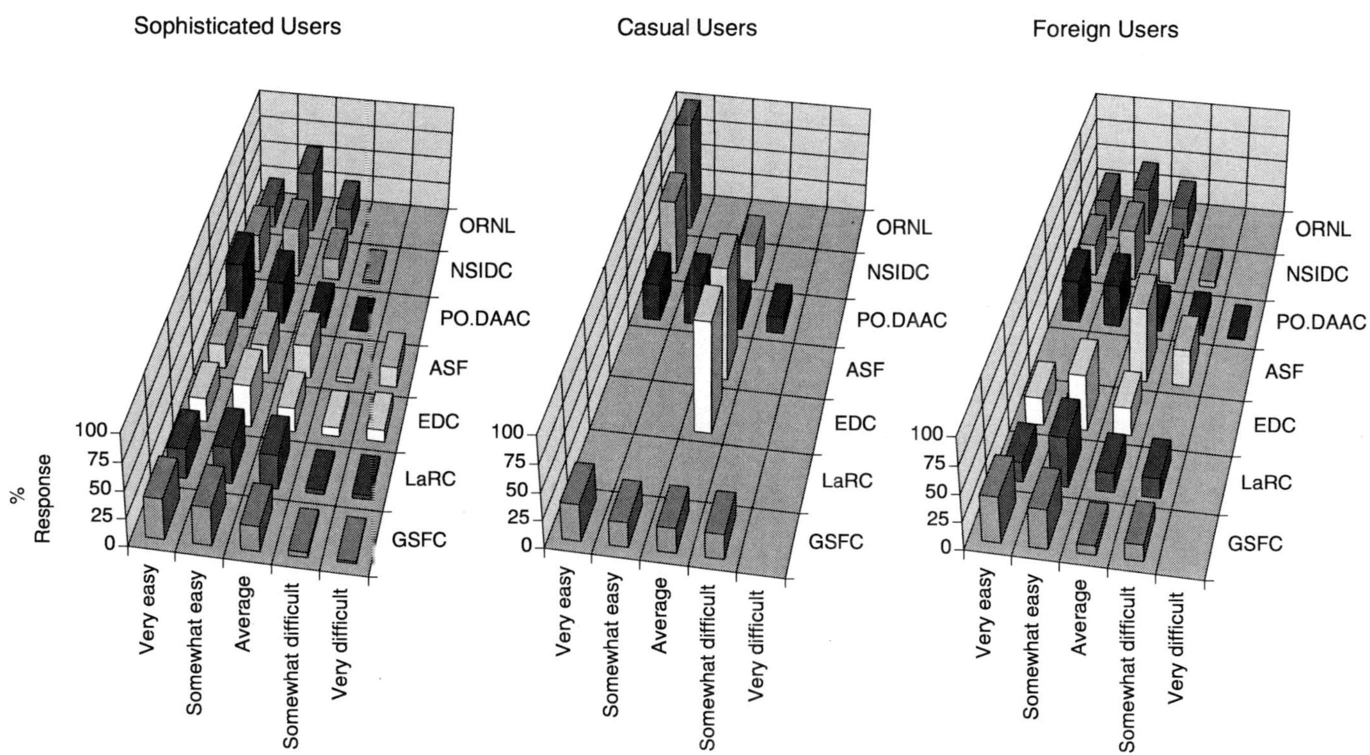

FIGURE 2.2. Ease of access to the DAACs. SOURCE: user survey (see Appendix D).

DAACs. The User Working Groups of the other DAACs have been less successful in defining or carrying out a useful role in DAAC operations.

The DAACs' main interaction with scientists, however, is with the EOS investigators (science and instrument teams) involved with the mission, as well as the broader scientific user community. The DAACs have had mixed success in their interaction with these two constituencies. Although the relationship between the DAACs and the EOS investigators was not explored in detail, the LaRC DAAC received strong praise from both its review panel and its data providers for becoming an integral part of radiation and cloud experiments. The other DAACs have been less successful in this regard, in part because EOS investigators are not always willing to involve the DAACs in their planning. In general, the DAACs judged by their review panels as being effective and successful centers are those that have achieved a level of symbiosis with the scientific user community they support. Particularly noteworthy from this point of view are the excellent reviews that the PO.DAAC and the NSIDC DAAC received. A better day-to-day interaction with the scientific user community, however, was recommended for the EDC, GSFC, and LaRC DAACs, especially since a strong scientific community with highly relevant interests resides within walking distance in all three instances!

The committee concurs with the panels' recommendations and indeed wishes to emphasize this point. The benefit of daily interaction with active researchers is immense, not only because it promotes data quality assurance, improves user services, and encourages innovations in response to research progress, but also because it makes the whole endeavor more responsive to the scientific demands for which it was undertaken. Although this conclusion is difficult to quantify, the committee feels strongly that it is an essential one and that to take action on this recommendation would lead to substantial long-term improvement of the DAACs.

A common theme echoed by all the DAACs is that it is extremely difficult to document the usage of their holdings in scientific publications. Even though most DAACs suggest a standard acknowledgment to be placed in scientific papers and ask scientists to send reprints of their papers, producing a comprehensive list of peer-reviewed science papers that make use of DAAC holdings is a daunting task. This is a problem faced by most data centers as well. The committee suggests that the DAACs post a list of publications (admittedly incomplete) that use DAAC data, directly on their respective Web sites, in a format that would permit an easy search for references with standard Web tools. The NSIDC DAAC's impressive list of 300 publications that use DAAC data from 1993 to 1997 would serve as a good model for other DAACs and data centers. Such a resource would be convenient to researchers seeking access to other publications based on DAAC data and would also encourage researchers to send relevant reprints to the DAAC as a way of "advertising" their own work. It also has the potential to become an excellent tool to help the DAACs document fully their role in the scientific enterprise.

Conclusions and Recommendations

With a few notable exceptions, users are satisfied with the performance of the DAACs. This assessment is based on the committee's user survey and the reports of the review panels, most of which conclude that the DAAC is doing a good job overall. In cases where the DAAC is having difficulty satisfying its users, the panels identify areas that could be improved. For example, all the panels recommended that the DAACs better characterize their user communities. This will help them to provide effective tools and services to established users, and also to increase the size and diversity of their user base. Promoting new uses for the data, which will arise inevitably with new user groups, will increase the return on the taxpayers' investment in EOSDIS. In addition, most panels noted that the DAACs need to improve their relationship with the scientific user community. Like the world and national data centers that the committee reviews regularly, those DAACs that have achieved a high degree of symbiosis with a collocated group of active researchers enjoy a substantially higher level of user satisfaction than others. As illustrated at the PO.DAAC and the NSIDC DAAC, such symbiosis provides advantages to both the DAACs and their scientific users, and can be accomplished through a variety of mechanisms (e.g., interaction with scientists on-site). Such collaboration increases the likelihood that the scientific objectives of the Earth Science Enterprise will be achieved.

Recommendation 4. In order better to track the rapidly growing and evolving population of EOSDIS users and serve the needs of existing and potential users, each DAAC should devise and implement quantitative measures for characterizing its current user community.

Recommendation 5. To function optimally, the DAACs need to be intimately involved with the scientific community they serve. The DAACs should deliberately pursue and improve routine, daily interactions with active scientists who use their data holdings. Among the many ways to meet this recommendation, the committee suggests (1) implementing a visiting scientist program; (2) encouraging DAAC personnel to pursue research endeavors, with the purpose of publishing the results; and (3) actively working with researchers within their host institution.

DATA MANAGEMENT: SOUP TO NUTS

Life-Cycle Data Management

The value of data collected through the EOS program lies not only in the short-term scientific gains that will arise from use of the newly collected data, but also in the long-term studies of changes in the global environment. Consequently, the notion of managing scientific data should not be reduced to a set of bureaucratically defined tasks. Life-cycle data management involves looking beyond immediate goals and deliverables and taking steps now in the interest of future generations of scientists and citizens to enhance their ability to make effective use of the unique and irreplaceable records that the EOS program is collecting at considerable expense. It begins with good instrument design and careful calibration, and extends through careful documentation of every step of the data processing and product generation, to reliable long-term archive.

The collections must be designed to convey the scientific and operational context of the measurements and all other ancillary information, that may assist in their proper interpretation at a time when none of the individuals originally responsible are available for questions. Past experience shows that such design is difficult and is frequently neglected, particularly because the science questions being asked change unpredictably with time. However, a consequence of such neglect is that the value of the archive for future generations of scientists is greatly diminished.

A life-cycle data management strategy is paramount when it comes to documenting long-term phenomena such as global environmental change. In the committee's view, data centers (and DAACs) should participate actively in all major stages in the life cycle of a data set:

- **Data collection**. The credibility and reliability of the data and data products depend on careful attention to calibration, internal consistency, and version control. Consequently, instrument teams and supporting processing staff place a high priority on collecting and recording this information. Much more difficult is capturing information deemed immaterial or common knowledge at the time of the experiment, or recording the strengths and weaknesses revealed by later use of the data set. Yet, this information may be critical to later interpretation of the data. The DAACs can help by participating in the planning of the data collection, seeking to clarify both the information that is being captured and the inputs or parameters that are imported from other sources, and to facilitate the process of recording them. Similar involvement with the metadata during the product generation stage is equally important. The LaRC DAAC has been successful in this regard, particularly in its participation in field experiments. For most ESE missions and experiments, however, the data collection environment is specified with little or no DAAC input.
- **Management of active data sets.** Active data sets require a curator who cares about the quality and completeness of the collection and has at least a general sense of its potential scientific value. Such a curator will (1) understand the strategic data needs of the scientific constituency; (2) prepare guide information to assist users in evaluating the relevance of the data to their purposes; (3)

develop tools and services, such as subsetting capabilities and Web interfaces, to help users find and work with the data; (4) contact experts on behalf of users with complex scientific queries; and (5) reprocess data in response to scientific demands. The latter is exemplified by the PO.DAAC, which reprocesses data regularly to keep up with scientific advances. Also noteworthy are the pre-subsetted data sets prepared by the GSFC DAAC, which enable users to work with manageable amounts of data, and the tools for documenting data sets provided by the ORNL DAAC. Most of the other DAACs, however, have to place greater attention on subsetting (LaRC and EDC DAACs), reprocessing (GSFC DAAC), or user services (EDC and ASF DAACs).

- **Long-term archive.** Archive of a valuable data set involves more than long-term storage. The assembly and presentation of useful information about a data set is equally important to its preservation for future generations. By planning for long-term archive, the DAACs ensure a greater likelihood that the data will remain useful beyond the period where a high volume of exchange, traffic, access, and manipulations takes place. Although the DAACs will not be responsible for EOS data collected more than 15 years past the end of the mission, their understanding of the data sets in their charge should be incorporated into the data sets before these are moved to a permanent archive. This will increase the probability that the holdings will retain scientific value.

The ECS metadata model lacks flexibility and is not extensible to permit adequate content-based access. In particular, better levels of spatial queries must be supported if the system is to fulfill the simultaneous goals of providing better access to information in general and supporting content-based access and subsetting of information in particular. For this purpose, convenient and efficient geospatial access for scientific images and data sets is essential. This requires effective, generalized gazetteer services, together with the ability to visualize spatial footprints for items in the collections. It is the combination of these two sets of services, together with the adoption of standards for representing geospatial metadata, that will significantly increase most users' ability to access the data they need without extracting large volumes of unneeded data from the archive. Several DAACs reported that they have raised these issues with ESDIS (and the ECS contractor) but have not prevailed, even though the matter of a better, more flexible, and extensible metadata model is fundamental to support scientific research (see "*The Role of NASA*," below).

Long-term custody of data is a difficult issue for NASA. NASA does not have a long-term archive mission, and it is not willing to shift funds from the EOS program to an archive agency such as NOAA. Nevertheless, NASA has concluded a Memorandum of Understanding (MOU) with the USGS, which now has a line item in its budget for the eventual acquisition of EDC DAAC holdings. MOUs are still being negotiated with the Department of Energy (DOE) for ORNL data and with NOAA for the remainder of the DAAC holdings. In practice, however, there is no coherent plan derived from a vision of scientific needs (insofar as these needs have been identified) for long-term archive of the data.

Any plan for long-term archive should consider both the scientific cost and the dollar cost of moving data sets to a different geographic location. The cost to the science comes from dissociating the data set from the scientists and data managers who have the relevant expertise to manage it. It is unlikely that staff at a scientifically unrelated data center, no matter how well intentioned, could manage the data as well. Additional scientific costs arise from the loss of data, which is inevitable in large-scale data transfers. The dollar cost of moving data sets from active archive to long-term storage is poorly quantified, but examples exist that could be used to calibrate the cost of such tasks. For example, the holdings of the now defunct Marshall Space Flight Center DAAC were transferred to several DAACs; the Spaceborne Imaging Radar-C (SIR-C) data set was exported—together with the processing hardware and software—from JPL to the EDC DAAC; and the Ocean Topography Experiment (TOPEX) data set was moved from the principal investigators to the PO.DAAC.

A related issue is the cost of adapting to constantly evolving technology, especially storage media and associated hardware. Given the pace of technological change and the DAACs' strategy of remaining within the technological mainstream, the DAACs will have to address this issue in their mid- and long-term plans. Yet, only the PO.DAAC reports such plans, perhaps because its restoration of SeaSat data illustrated the difficulties that arise from gaps in the metadata, outdated storage media, and machine-specific data formats. Its excellent document, "SeaSat Data Restoration - Lessons Learned" that it produced for this review should serve as a valuable resource to other DAACs and data centers.

Finally, it is well known among data center managers that 10% of the users access 90% of the holdings, whereas 90% of the users access 10% of the holdings (generally at a higher level of data processing). This causes a dilemma for the DAACs, particularly with the enormous volumes of data they will face. Ten percent of the users will require advanced data storage and distribution capabilities for dealing with enormous data sets, and ninety percent of the users will need much smaller, higher-level, interpreted data sets on more accessible media (i.e., CD-ROMs). The DAACs therefore must resolve one or both of the following issues: (1) subsetting large data sets into manageable size in a sufficiently short time to satisfy on-line data requests of a large fraction of users and (2) identifying and implementing means of distributing very large data sets to a small fraction of users. The DAACs should plan to satisfy both requirements.

Strategic Planning

To be successful, each DAAC must create a vision of what it wants to accomplish and a strategy for achieving that vision. The vision should go beyond a simple statement of ESDIS requirements such as handling the EOS data flow and producing the planned products. It should influence every aspect of DAAC operations, including participation in the flight missions and experiments, acquisition of data sets, service to an expanding and evolving user community, and accommodation of the rapid evolution in computing, storage, and communications technologies. Similarly, the system-wide Strategic Management Plan produced each year by ESDIS and the DAACs (see Appendix C) should take into consideration the need for the DAACs to act as a coherent system. Once in place, the DAACs will also have to develop quantitative metrics to monitor the performance of the process or the system as a whole. Few of the DAACs, however, have engaged in this thought process, and most of the panels recommend some level of strategic planning. (Notable exceptions are the PO.DAAC and the NSIDC DAAC.) For example, a vision and an implementation plan would help the EDC DAAC to serve the needs of its scientific and, potentially much more numerous applications users, and would help the ORNL DAAC to become involved in EOS flight programs. The latter is particularly important because the unique biogeochemical holdings of the ORNL DAAC are essential to the proper validation and calibration of remotely sensed data. Unless the ORNL DAAC asserts itself, the various flight projects will be forced to develop independent solutions, and large components of the EOS program will fail.

The panels' most common recommendation in this regard, however, has to do with technology. Strategies for acquiring and upgrading hardware and software are difficult to develop because such tasks are partly the responsibility of the ECS contractor. This issue is particularly important for the AM-1 DAACs, which have received significant amounts of ECS hardware and software (often at their own request) but know little about what will be delivered in the future. Much of this equipment is literally sitting on the computer room floor awaiting the anticipated huge data flows and is slowly becoming rather less than state of the art even before really being used. The system will thus employ multiterabyte technology when the most sophisticated users will expect petabyte capabilities.

Although the DAACs cannot control completely the type of ECS equipment they will receive or its schedule for delivery, they can choose their own hardware and software for managing existing data sets. (The GSFC and LaRC DAACs also chose their own systems for managing data from the TRMM mission.) When acquiring hardware, most DAACs aim to stay within the technological mainstream, a standard goal for data centers because limitations on financial resources prevent much experimentation with technology. Moreover, current wisdom about the evolution of technology (i.e., "Moore's law") is that capacity—processor speed, bandwidth, mass storage—now doubles every 18 months. This time constant is comparable to that of the procurement cycle, which makes it difficult for the DAACs to adopt an agile response to the acquisition of new technologies. The committee notes that this problem is particularly acute with the ECS, which chooses equipment far (sometimes years) before deployment. The panels report that the equipment purchased by the DAACs generally falls within the technological mainstream, although some development efforts are based on technologies that are more advanced than those included in the ECS. For example, the transition to the next generation of processors will likely place the LaTIS software at the LaRC DAAC ahead of the ECS and make it incompatible with planned ECS distributions, or so-called drops. Thus, any strategy for the acquisition and upgrade of hardware and software will also have to take into account compatibility, not only with the ECS, but also with the other DAACs in the system.

A related aspect is that the DAAC system—perhaps overwhelmed by the prospect of facing the huge EOS data flow—has spent little time selecting measurable goals and self-assessment criteria against which to gauge collective performance. Without quantitative measures of performance, it will be difficult for NASA or the DAACs to determine whether the needs of all the EOSDIS constituencies are met and, thus, whether the DAAC system as a whole is a success. At the moment, launch delays and ECS failures have tarnished the image of all components of EOSDIS, including the DAACs. Yet, as pointed out by the panels, most of the DAACs function well individually, even though the same cannot be said of their behavior as a system. A regular peer review that focuses on established performance measures for the DAACs and for the DAAC system would help provide a sound basis for determining the health of the DAACs.

Conclusions and Recommendations

The ultimate success of the Earth Science Enterprise will be judged not only by the immediate scientific gains that arise from use of newly collected data, but also by the ability of scientists to use the data in the long term to study global environmental change. The data are most likely to retain their usefulness in the long term if the DAACs adopt a life-cycle data management approach and become involved in everything from the collection of data to its eventual archive. The PO.DAAC has such a comprehensive management philosophy, but most DAACs focus on the middle step—management of active data sets—and they do so successfully overall. Only a few DAACs participate in the design of the data collection environment. The remainder either do not see such involvement as one of their roles or are discouraged from participating by the science and instrument teams. Both sides will have to come to the table if a mutually beneficial relationship is to develop. Finally, the committee notes that there is still no concrete plan for the long-term archive of the vast majority of DAAC data. Because NASA has no archive mission, the DAACs will have only a limited role in the transition of DAAC data sets to archives of other agencies. Nevertheless, their knowledge of the data and experience with data transfers should prove valuable in this process.

Implementing a life-cycle approach to data management requires a greater degree of planning than currently exists at most of the DAACs. Only a few

DAACs have a clear vision of what they are trying to achieve or strategies for achieving their goals. In addition, all the DAACs could benefit from devising and monitoring quantitative performance measures. Such measures are useful for evaluating the performance of a function or process, as well as for determining the success of a DAAC or the entire DAAC system.

Recommendation 6. Ongoing changes in data volumes, user expectations, and emerging technologies are powerful forces that put pressure on each DAAC to evolve independently of the others. In order to counteract such centrifugal forces, each DAAC should prepare and periodically update a practical strategic plan for dealing with change, while preserving the concept of a coherent system.

Recommendation 7. Excellence in a research enterprise is best gauged through assessment of performance by one's peers, according to a commonly accepted set of performance criteria. The DAACs must develop a set of quantitative performance metrics by which they can measure their own progress and evaluate their success as individual centers and as a coherent system. Periodic peer review aimed at gauging accomplishments against these metrics should be incorporated as part of this ongoing process.

The committee emphasizes that Recommendation 7 is aimed at the DAACs, rather than the ESDIS Project. This is because such peer reviews should not be a bureaucratic imposition, but a means by which the DAACs and their user communities achieve a greater understanding of one another and thus a more effective level of service.

THE ROLE OF NASA

As originally envisioned by its creators, EOSDIS serves three important roles. First, it provides a mechanism for distributing data from EOS-related missions and experiments. Second, it promotes creative scientific analysis of these data and, as such, must enhance opportunities for scientists to build on the unprecedented information it already contains and on the new data anticipated over the coming decades. Third, it is the largest single component in global efforts to understand, predict, document, and mitigate the impacts of global environmental change (although key data currently reside in other agencies such as NOAA.) Consequently, EOSDIS will greatly influence complementary efforts in other agencies and throughout the world. Strong leadership within NASA is critical for implementing this vision and for balancing the potentially conflicting demands of the constituent elements of EOSDIS—the ECS, DAACs, EOS investigators, and users.

Since EOSDIS was designed, the Internet and the World Wide Web, along with greatly reduced costs of computation, have changed the paradigm for scientific collaboration. Scientists no longer need to rely on a centralized warehouse of data equipped to respond to a sharply limited range of predefined queries. Instead, distributed databases and information systems offer the possibility of finding useful information in much more flexible ways. Probably the greatest challenge facing EOSDIS at this time is to adjust its social and management structures to take full advantage of these new opportunities without jeopardizing its ability to manage high-volume data streams from coordinated instrument systems.

Changing established ways of doing things is generally painful and involves some risk. It also requires dedicated leaders with initiative and vision. For EOSDIS to succeed in this new environment, its constituent elements must be empowered to fulfill their special roles. For example, the DAACs (or relevant Earth Science Information Partners) should be vested with the appropriate authority to take all actions necessary to satisfy the needs of the science community. This is particularly true for the ASF DAAC, which lacks authority to compel JPL to develop an information system that is responsive to the needs of the DAAC or its users. Each center should be encouraged to develop its own personality in accordance with its special needs, as long as it is contributing to the evolution of a responsive and dynamic distributed system for the EOSDIS community at large. Neither the ORNL nor the ASF DAAC, for example, fits the "standard" DAAC mold because they don't deal with data from EOS spacecraft. Yet each has a vital role to play in the Earth Science Enterprise and should be integrated conceptually into EOSDIS. (It is significant that NASA's model of the EOSDIS architecture divides the ORNL DAAC from the others [see Figure 1.1].) In the case of the ASF DAAC, NASA will have to first develop a long-term policy on the acquisition, processing, and use of SAR data for civilian purposes.

ESDIS will have to create incentives for making the constituents respond to the needs of the broader community and safeguards for ensuring that they do not destroy the overall integrity of the data system. This delicate balance requires vision and leadership from both the DAACs and ESDIS. In fact, the ability to exercise leadership in this regard should be a substantial consideration in personnel selection.

Finally, metrics of success must be devised and agreed to, that recognize that the value of EOSDIS lies in the scientific understanding and reliable information emerging from the entire program, rather than in the cost per byte of data processed. These are not trivial requirements for NASA management, and effective strategic planning requires leadership from the highest levels.

Conclusions and Recommendations

Under the new EOSDIS model, ESDIS will have to forgo the current mode of operations in which all strategic decisions are made centrally and adopt the mode of providing incentives to individual DAACs (1) to serve their individual specialized scientific constituencies most effectively and (2) to collaborate to provide users with a common look and feel to the information system. The former will require close collaboration with science and instrument teams and active scientists. The latter will involve some compromises with respect to the ideal of one-stop shopping. In the right governance structure, the DAACs would task ESDIS and ECS to address common needs, which must be identified from the bottom up rather than from the top down. Such common needs range from technical issues, such as format translations and regridding data in different projections, to high-level issues, such as the completeness and extensibility of the metadata model and what kind of services to provide to users. In effect, ESDIS will have to foster the creation of a federation of DAACs by delegating some of its authority for serving users to the DAACs. Such a federation differs from NASA's prototype federation (see Chapter 1, "*Federation and Recertification*") in that the partners require greater stability and continuity. In addition, because of product interdependence, the DAACs have a requirement for reliability and adherence to schedules. In this sense a DAAC federation would constitute a core around which the broader federation of (sometimes ephemeral) ESIPs could grow and function effectively.

> **Recommendation 8. The DAAC-ECS-ESDIS model for managing EOSDIS data and information has not succeeded. To take advantage of new technological approaches and management models, ESDIS should foster the creation of a federation of DAACs by delegating to the DAACs some of its authority for serving users and by providing incentives to the DAACs to serve the broader community as well as their individual specific constituencies.**

OTHER ISSUES

The Cost of the DAAC System

The EOSDIS budget is currently about $2 billion over a 10-year period. The DAACs, including ECS-provided hardware, software, and personnel, account for about 30% of the EOSDIS budget, or $60 million per year. The remainder of the EOSDIS budget provides for data capture and communications, ECS development, science computing facilities, and program management.

It is important to note that the DAAC budget figures provided by ESDIS and presented in this report are only approximations. The true cost of the DAACs is difficult to determine for the following reasons:

- the DAACs receive funds, services, and personnel from several sources, including NASA Headquarters, the ESDIS Project, and their host institutions;

• resources are shared (i.e., the ECS development effort benefits several of the DAACs);

• neither NASA nor the DAACs (except ORNL) practice full-cost accounting; and

• congressional appropriations for the EOS program are commonly less than NASA's request.

It is also important to note that the budget histories presented in the DAAC chapters were provided by ESDIS in May 1998. The values for FY 1994 to FY 1997 are actual values, but values for FY 1998 to FY 2002 are projections, which are likely to change significantly as funding for EOSDIS declines and backup plans are implemented. Because the numbers were determined by the DAAC's primary funding source (ESDIS), they provide the most complete and consistent picture available of the cost of the DAAC system. The cost estimate provided by ESDIS is significantly higher than the cost estimates provided by most of the DAACs because it includes the ECS-provided hardware, software, and personnel deployed at the DAACs. The DAACs have little say about these resources and tend to consider them ECS, rather than DAAC, expenses. The committee and its panels, however, believe that all resources at the DAACs should be included in cost estimates of the DAAC system, so both DAAC and ECS-related expenses at the DAACs are itemized in the DAAC budgets (see Chapters 3 through 9).

However, even with the ECS costs factored in, it is clear that the DAAC system is less expensive than is commonly believed. For example, the DAAC budget was presented misleadingly as $100 million per year at a 1995 NRC workshop in La Jolla, California. The apparent high cost of the DAACs was one of the drivers for proposing a federation management model for EOSDIS. Several survey respondents also commented that the DAACs are too expensive.

All of the panels attempted to assess the cost-effectiveness of the DAACs, but several DAACs could not even suggest suitable metrics. (The PO.DAAC and the ASF and GSFC DAACs measure cost-effectiveness as the unit cost of delivering a data set; the ORNL DAAC measures it as the cost per unit of data stored.) Consequently, the issue of cost-effectiveness could not be addressed in a significant way. It is noteworthy, however, that even though most of the DAAC budgets far exceed the budgets of national data centers, none of the national data center directors who served on the panels thought that the DAAC budgets were too high for the amount or complexity of data being handled, the size of the user base, or the services provided.

Impact of Contingency Plans

Because of delays in the ECS, both the DAACs and the science and instrument teams have developed fall-back strategies for processing and disseminating

EOS data and data products. Implementing these contingency plans will result in some data and data products becoming available, but not as many as had been hoped under the original EOSDIS design. For even a reduced number of products to be distributed, ESDIS, the DAACs, and the various teams will have to resolve the following issues:

- **Documentation**. The DAACs are likely to play a much smaller role in generating products and will therefore be less knowledgeable about the data sets and data products in their charge. Consequently, documenting what is being done to the data in the product generation stage becomes even more important under the contingency plans than it was before. If the DAACs are to provide an adequate level of user services, they will need to have a much closer relationship with the science and instrument teams than currently exists. In particular, they will have to become more involved in producing the metadata for the data products.
- **Coordination**. Under the 25-50-75 scenario, only 25% of the data will be made available initially. It is up to the individual science team to decide which 25% of the data will be processed. This decision will affect not only users of that data product but also other science and instrument teams and DAACs that need to use the data to produce other data products. To accommodate product interdependencies, the DAACs and teams both will have to place a high priority on coordinating schedules. Otherwise, the production of many important data products is likely to be further delayed.
- **Dissemination**. In the earlier stages of data processing and distribution, the DAACs are likely to be bypassed in favor of scientists calling their instrument team colleagues. In fact, the Science Computing Facilities where the data products are being generated are likely to become the primary distribution mechanism until the instrument teams members become so fed up that they relinquish the distribution task to the DAACs. (The DAACs serve tens of thousands of users each year [see Table 1.3]). As the ORNL DAAC can attest, awaiting data sets that could arrive from principal investigators at any time is frustrating and makes it difficult to allocate personnel and computer resources for distributing the data to the broader community.

It is unclear whether one-stop shopping is still a realistic goal for EOSDIS as a whole. Evolving Web access tools will increasingly permit reliable distributed searches, but only if there is a common terminology of keywords that is also shared by users. The DAACs and their science constituencies together have to develop this terminology. If this is done, it should also be possible to make available complete granules from standard products in a manner that is reasonably consistent overall. Because the granules typical of large, low-level data sets are generally too large to download over the Internet, and because subsetting tools designed by the ECS will not be ready in time for the AM-1 launch, the DAACs most affected will have to incorporate subsetting into their contingency plans.

CONCLUSIONS

With technical problems in the EOSDIS Core System and, more recently, flight operations commanding the attention of NASA and Congress, it is easy to lose sight of what the EOS program is all about—understanding the Earth and the processes that govern it. Because such a wide variety of data will be collected—atmospheric, oceanic, polar, biospheric, and solid earth—scientists will be able to use EOSDIS to support both disciplinary and multidisciplinary research. Although multidisciplinary scientists are only one component of the EOSDIS user community, meeting their needs will be the greatest challenge of the system. For these users, EOSDIS must be more than a collection of discipline data centers; it must be a real system that enables users to access and combine data from more than one DAAC or data center.

The current collection of DAACs does not as yet function as a system. To become a system in reality, the DAACs and ESDIS will need a common vision of the goals of the system and a commitment to developing practical approaches toward achieving these goals. Developing such a system becomes an even greater challenge as EOSDIS evolves to a more distributed federation. NASA leadership is crucial for this transformation to succeed. In the near term, NASA's attention is necessarily focused on fulfilling existing software commitments and on supporting science and instrument teams for existing or near-term flight missions. However, in the longer term, EOSDIS must establish by force of example its role as the creative nerve center for the scientific understanding of the Earth in the next decade. Though in many respects still a dream, this goal is too important to let slip away.

3

Goddard Space Flight Center DAAC

Panel Membership

J.-BERNARD MINSTER, *Chair*, Scripps Institution of Oceanography, La Jolla, California

FERRIS WEBSTER, *Vice Chair*, University of Delaware, Lewes

SYDNEY LEVITUS, NOAA National Oceanographic Data Center, Silver Spring, Maryland

RICHARD S. LINDZEN, Massachusetts Institute of Technology, Cambridge

TERENCE R. SMITH, University of California, Santa Barbara

JOHN R.G. TOWNSHEND, University of Maryland, College Park

ABSTRACT

The Goddard Space Flight Center (GSFC) DAAC is the largest of the EOSDIS DAACs. It manages a variety of data sets related to climate, the biosphere, and the upper atmosphere, and it will also process, disseminate, and archive data from the flagship EOS instrument, the Moderate Resolution Imaging Spectroradiometer (MODIS). The DAAC understands its role and is doing a good job with its current data sets. However, the large data volumes and complex algorithms of the MODIS data stream present a significant challenge to the DAAC, and the panel's main recommendation is that the DAAC continue to focus its efforts on preparing for the AM-1 platform, and particularly the MODIS instrument.

INTRODUCTION

The GSFC DAAC was created in 1993 to archive and distribute data related to climate change, atmospheric dynamics, global biosphere, hydrology, and upper atmospheric chemistry (Box 3.1). Its roots are in the NASA Climate Data System and the Pilot Land Data System. The first data sets archived by the DAAC included data collected by the Total Ozone Mapping Spectrometer (TOMS) and the Nimbus-7 Coastal Zone Color Scanner (CZCS). Today the DAAC manages data sets from a variety of missions and experiments, supports the Goddard Data Assimilation Office, and also manages some of the hydrology holdings of the Marshall Space Flight Center DAAC, which was closed in 1997. With a staff of 114 and current holdings of 4 TB, the GSFC DAAC is one of the largest DAACs in the EOSDIS system.

In the EOS AM-1 era, DAAC holdings will increase in size by a factor of 500 (Box 3.1). The Sea-Viewing Wide-Field-of-View Sensor (SeaWiFS) and Tropical Rainfall Measuring Mission (TRMM) instruments, which have already been launched, will produce 65 TB of data, and MODIS, which will be launched in early 1999, will produce nearly 2,000 TB. To prepare for these large data streams, the DAAC is staffing up. Approximately 40 EOSDIS Core System (ECS) contractors have been added to process MODIS data, and about 12 permanent staff have been added to manage DAAC operations. The average budget for the DAAC, which includes DAAC personnel and functions, civil servants, ECS contractors, and ECS-supplied hardware, is about $15 million per year.

Managing the enormous MODIS data stream poses daunting managerial and technological challenges for the GSFC DAAC. Of most concern is whether the information system, particularly the ingest system, can be scaled up to accommodate increasing loads (see "*Technology*," below). To prepare for the new data streams, the DAAC will start "day-in-the-life" exercises and operations rehearsals several months before launch. As of June 1998, the ECS was still not ready for day-in-the-life exercises, but so far, it has been sufficient to test the science algorithms. Delays in the launch of the EOS satellites will provide additional preparation time.

The Panel to Review the GSFC DAAC held its formal site visit on October 20-21, 1997. To ensure that its report and recommendations reflect recent developments, several panel members visited the DAAC again in June 1998. The following report is based on findings from both visits and e-mail discussions with DAAC managers in July and September 1998.

HOLDINGS

Even before the launch of TRMM and AM-1, the GSFC DAAC has been managing and distributing numerous data sets of substantial size. These include in particular the Advanced Very High-Resolution Radiometer (AVHRR) and the

BOX 3.1. Vital Statistics of the GSFC DAAC

History. The GSFC DAAC was created in 1993 out of the NASA Climate Data System and the Pilot Land Data System. Its holdings go back to 1978.

Host Institution. NASA Goddard Space Flight Center in Greenbelt, Maryland.

Disciplines Served. Atmospheric science and hydrology; data are available on the chemistry of the upper atmosphere, global biosphere, atmospheric dynamics, and climatology.

Mission. To maximize NASA's investment benefits by providing data and services that enable its customers to fully realize the scientific and educational potential of data and information from the Earth Science Enterprise.

Holdings. The DAAC holds 4 TB of heritage data sets and anticipates receiving more than 2000 TB of data from the AM-1 platform.

Users. There were 12,216 unique users in FY 1997, based on log-in addresses.

Staff. In FY 1998 the DAAC had 74 staff (9 of them civil servants) and 40 ECS contractors.

Budget. Approximately $9.2 million in FY 1998 (including DAAC costs and ECS-provided hardware, software, and personnel), increasing to $17 million in FY 2000.

Television and Infrared Observation Satellite Operational Vertical Sounder (TOVS) Pathfinder data sets, which have been used extensively by EOS investigators to prepare for the processing of AM-1 data (Box 3.2). These holdings consist primarily of imagery and remotely sensed data, and constitute one of the best resources available to date to support research on the atmosphere and global climate change. Figure 3.1, for example, illustrates changes in the size of the ozone hole, as detected by several remote sensing instruments, whose data are managed by the GSFC DAAC.

BOX 3.2. Data Holdings as of January 1998

- *Total Ozone Mapping Spectrometer (TOMS)*—Data from the Nimbus-7 and Meteor-3 satellites for November 1978 to December 1994.
- *Upper Atmosphere Research Satellite (UARS)*—Products from nine instruments for September 1991 to present.
- *Television and Infrared Observation Satellite (TIROS) Operational Vertical Sounder (TOVS)*—1-degree resolution data for 1978 to 1994.
- *Sea-Viewing Wide-Field-of-View Sensor (SeaWiFS)*—Data now available provide local, regional, and global coverage.
- *Greenhouse Effect Detection Experiment (GEDEX)*—Global, regional, or local data sets for the 1980s.
- *International Satellite Land Surface Climatology Project (ISLSCP) Initiative I: Global Data Sets for Land-Atmosphere Models*—Monthly, monthly-six-hourly, and six-hourly data are available globally on a 1-degree grid for 1987 to 1988.
- *Nimbus-7 Coastal Zone Color Scanner (CZCS)*—Data at 1-km, 4-km, or 20-km resolution for November 1978 to June 1986.
- *Pathfinder Advanced Very High-Resolution Radiometer (AVHRR)*—8-km-resolution data for 1981 to 1994.
- *Goddard Data Assimilation Office (DAO)*—2- × 2.5-degree-resolution data for 1985 to 1993.
- *Moderate Resolution Imaging Spectroradiometer (MODIS) Airborne Simulator (MAS)*—Data from nine campaigns and other data sets are available on tape.
- *Tropical Ocean Global Atmosphere-Coupled Ocean Atmosphere Response Experiment (TOGA-COARE)*—Data from surface, aircraft, and satellite measurements for November 1992 to February 1993.
- *Marshall DAAC hydrology data sets.*
- *Interdisciplinary Climatology Data Collection*—Monthly data for land, oceans, and atmosphere are available globally on a 1- × 1-degree grid.

SOURCE: NASA (1998).

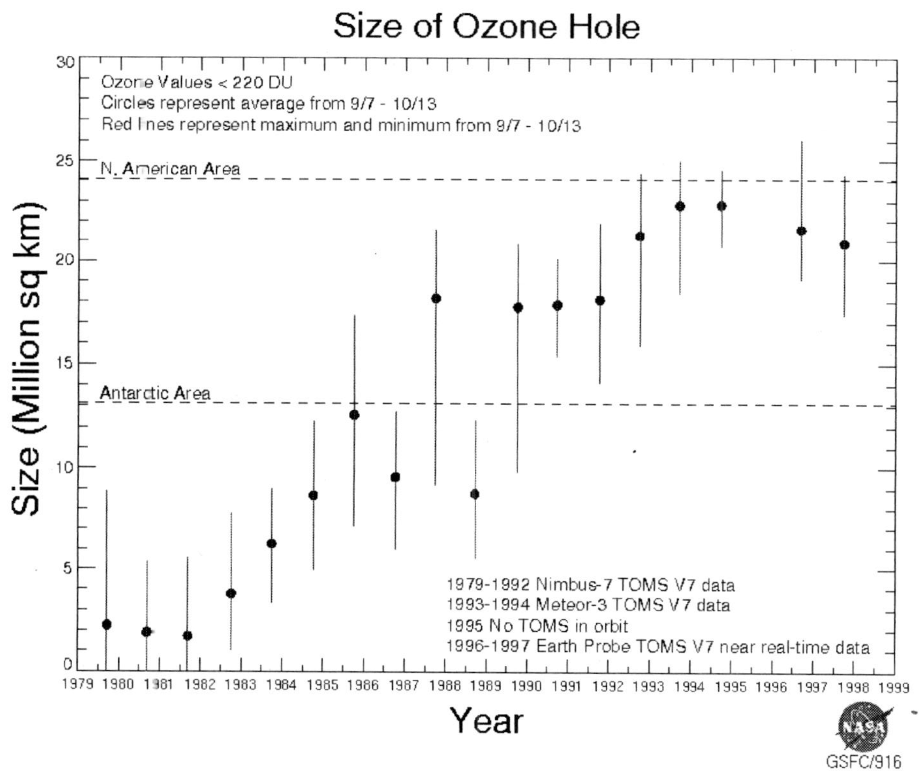

FIGURE 3.1. Change in the size of the ozone hole between 1979 and 1998, based on data from the Total Ozone Mapping Spectrometer (TOMS). Vertical lines represent the maximum and minimum from September 7 to October 13. SOURCE: Atmospheric Chemistry and Dynamics Branch of NASA Goddard Space Flight Center.

Metadata

The current metadata "model" is meager, and a much richer metadata model must be developed if the system is to migrate toward a greater degree of content-based access. In particular, the system must be able to support better levels of spatial query if it is to fulfill the goal of providing better access to information in general and content-based access and subsetting of information in particular. It is important that geospatial access be supported for scientific images and data sets. This requires good, generalized *gazetteer* services and the ability to represent *spatial footprints* for items in the collections. The combination of these two sets of services, together with the use of standards for representing such geospatial metadata, greatly increases a user's ability to access appropriate information. The current support for these services in the GSFC DAAC is still relatively primitive. The center, however, could make significant progress in this area by using resources developed elsewhere.

Processing Plans

At the time of the site visit in October 1997, the DAAC was to have been responsible for Level 0 ingest and archive, and Level 1 through 3 production and archive of MODIS products (see Table 1.1 for a description of processing levels). ESDIS is now considering funding the MODIS science teams to process the products (Level 2 and higher) because of concerns that the ECS will not be available in time for launch. Similarly, the MODIS instrument team may process MODIS Level 2 and higher land products and snow and ice products, which are currently scheduled to be processed at the EDC and NSIDC DAACs, respectively.

Until a final decision is made, the GSFC DAAC will continue testing the MODIS Product Generation Executables (PGEs) and integrating them into the processing system. As of June 1998, the DAAC had successfully integrated seven of the 40 PGEs into the ECS data processing system and had run an additional eight PGEs outside the ECS. In addition, a chain of three PGEs has been successfully tested and longer chain tests are being planned. On the other hand, if the Level 2 and higher products are to be processed by the MODIS science team, the GSFC DAAC will have to integrate only three PGEs into the system. If all the processing is to be done by the science team, the DAAC's role will be reduced to archive, dissemination, and user services.

Reprocessing Strategies

As research using DAAC data progresses, better algorithms will be devised by the scientists, and errors will likely be discovered in the data products. Ultimately a need will arise to reprocess the entire data set *ab initio*. Consequently, the DAAC must plan to allocate resources for reprocessing tasks. The science teams decide when reprocessing should be done, and the ESDIS Resource Allocation Board and the Science Review Committee arbitrate when several science projects request reprocessing. If the choice is between reprocessing an existing data set and processing data from a new instrument, the DAAC's priority is generally on the new mission. It appeared to the panel that the DAAC considers reprocessing an additional thing to do, rather than an integral part of its data management role. The panel is concerned about the consequences on the user community of the DAAC's decisions on what or when to reprocess.

Recommendation 1. The DAAC should incorporate reprocessing as an integral part of its data management strategy and plan for adequate resources over time for reprocessing needs arising from errors in product generation or algorithm improvements.

Subsetting Strategies

MODIS files will be so large that many users will not have the hardware, software, or personnel capability to produce subsets they can work with. Even though TRMM data sets may not be unmanageably large, the DAAC plans to make its data sets easier to obtain and use by preparing canned subsets in simple formats. The products are customized for particular disciplines or types of customers. The DAAC then advertises the packages and tries to get other customer groups to use them. Custom subsets for individuals are too expensive to produce, and the DAAC hopes to develop on-the-fly subsetting capability to meet their needs. The panel agreed that the production of canned subsets is a good strategy, and it encourages the development of on-the-fly subsetting.

Treatment of Model-Derived Data

When dealing with EOS data, it is crucial to distinguish between data products, which consist simply of the results of measurements, and derived data, which depend on specific model calculations. The latter include not only the meteorological variables derived by the Goddard Data Assimilation Office (DAO), but also all variables that depend on first guesses from the data assimilation model. Such derived data are dependent for their utility on the accuracy and reliability of the model. At present, the pressure of the AM-1 launch date has all but frozen efforts to correct errors in the DAO's GEOS model so that derived data products can be immediately available on launch. The panel believes that such time pressure might be appropriate to data stemming directly from AM-1 measurements, and is appropriate for activities such as archiving, but is not appropriate for derived products whose utility for climate studies depends on the adequacy of the underlying model.

Moreover, the panel's view is that the notion that climate data are needed in real time is a contradiction in terms. Climate data extend over sufficiently long periods to define climate. Under the circumstance, there is no meaningful time pressure to have a model regardless of its problems. For derived data, model performance is a critical component of data quality control, to which the GSFC DAAC has paid inadequate attention. The common practice of model validation in which one searches for similarities between model output and directly observed data is insufficient for this purpose. Rather a program of testing and evaluation is needed. This should be of as much concern to the GSFC DAAC as to the DAO. The panel felt that neither the DAAC nor the DAO liaison with the DAAC seemed to be sensitive to this issue. Under these circumstances, the DAAC seems to be part of a diffusion of responsibility that leaves no one accountable. There should be some mechanism whereby data products can be accepted by the DAAC for dissemination only if these products meet strict scientific criteria. So far, there is no evidence of such a mechanism. The panel encourages the DAAC to set up these procedures as soon as possible.

Long-Term Archive

NASA and NOAA are negotiating a Memorandum of Understanding for the long-term archive of EOS data sets. To help ensure the long-term vitality of the data, NASA has provided some funding to NOAA to prototype an archive. The prototype system is based on the GSFC DAAC's Version 0 system, but the DAAC has no role either in developing the prototype or in ensuring the long-term (the so-called 20-year test) usability of the data. Because the DAAC understands its data sets, the panel believes that it should become involved in the crucial process of transferring responsibility for the long-term archive at the earliest possible stage.

USERS

Characterization of the User Community

The DAAC does not have a well-defined user model. Instead, it divides "customers" into three levels of sophistication: (1) research scientists; (2) application users and college students; and (3) high school teachers and students. The first group—EOS instrument team members, EOS interdisciplinary science team members, and non-EOS investigators—includes both users and providers of data.

The DAAC does not seem to have a clear idea of which customers are its highest priority to serve, and no user community feels as if it "owns" the DAAC. Moreover, although the DAAC solicits input from its user communities via surveys and annual User Working Group and "voice-of-the-customer" meetings, there seems to be no systematic process by which the DAAC gains an improved understanding of what its priorities should be. To serve the needs of its user community effectively, the DAAC will have to constantly refine its understanding of the user community's characteristics and needs.

User Working Group

The User Working Group (UWG) membership is weighted heavily toward data providers. In the past, the UWG focused on issues such as guiding the development of the ECS and setting priorities on data sets. Now that most of these issues have become settled, the UWG is trying to define a new role for itself. Some members want the UWG to function like an external review panel that has clout with NASA, and others want it to continue functioning like an advisory panel composed predominantly of data providers.

The DAAC director implements most of the recommendations of the UWG and also uses the UWG to provide protection and endorsement of the market approach to data management (see "*Management*," below). The panel felt that less emphasis on the latter and more emphasis on critically reviewing DAAC activities would improve the effectiveness of the UWG.

Interaction with the Scientific Community

Although the DAAC is customer oriented, its relationship with its primary user community, the scientific community, is substantially weaker than it should be. The DAAC is situated within a large research facility, but there is little interaction with ESE scientists at Goddard Space Flight Center or elsewhere. In fact, the absence of on-site scientists is seen as an advantage by the DAAC because it gives the DAAC independence. Consequently, there appears to be no mechanism for scientists to provide feedback. DAAC staff, including those hired from the Earth Science Division at Goddard, are generally not carrying out research using the data sets, and it is not clear that they maintain a working relationship with those who are. Hence when users approach the DAAC about the data sets the DAAC distributes, it is not clear that they will get appropriate high-level responses about the quality of the data. There seems to be no well-structured process that would allow "complex" inquiries to be passed quickly to those who had generated and/or were using the data sets.

Recommendation 2. Interactions between the DAAC and the science community should be improved. Examples of actions the DAAC might consider include (1) establishing a visiting scientist program, (2) hiring a full-time DAAC scientist, and (3) collocating DAAC staff with Goddard Space Flight Center researchers.

User Services

Beginning in 1994, the User Services Group was abolished and the DAAC was reorganized into a Customer Support Group and an Engineering Group. The Customer Support Group—made up of several data support teams—provides end-to-end support for specific scientific disciplines, and each team is responsible for data preparation, information management, building the customer base, and interacting with customers for an individual science discipline. The teams have the flexibility to determine how to carry out their responsibilities. The panel was impressed that the data teams are so empowered and that the organizational changes have apparently led to speedier resolution of customer problems.

Foreign Access

The GSFC DAAC is located within a secure facility. Although the Internet provides a means of visiting the DAAC virtually, foreign visitors face difficulties in obtaining clearance for a physical visit of more than a few days.

Usability of Data Within a Scientific Context

Earth scientists will inevitably need access to NOAA as well as NASA data. Some of the necessary NOAA data are difficult to access, and this will ultimately inhibit the utilization of NASA data. The two cannot really be decoupled in a scientific context. Fortunately, the relative volume of NOAA data in the EOS context is small; it is also likely that at least the NOAA data in digital form could be incorporated at modest cost. Even the digitization of existing paper data should be affordable. NASA should consider assisting NOAA in this regard since the inclusion of these data in the DAAC will add greatly to its value as a climate research resource.

TECHNOLOGY

Strategy

The DAAC's hardware strategy is to stay on the leading edge of mainstream. It allows others to test and find the bugs in new technologies, then it adopts the technologies that work. The DAAC manager is always thinking ahead, but no real plans are made more than six months in advance; this allows the DAAC to retain its flexibility. However, the panel feels that this tactic limits the DAAC's ability to deal with certain medium- to long-term issues: the rule of thumb for doubling performance in computer technology (so-called Moore's law) is only about 18 months, so that the DAAC will likely have to face retooling several times over its lifetime. At the same time, the useful life of ECS-provided hardware and software is pegged nominally at five years, making it difficult for a production facility to evolve in response to continuing technological advances. Consequently, the panel feels that it would be prudent for the DAAC technology team to have a more specific strategy for tracking developments in relevant areas of software, hardware, and communications.

Recommendation 3. In order to stay on the leading edge of mainstream, the DAAC should formulate a long-term strategy for keeping up with and taking advantage of new technologies.

Hardware Availability

The DAAC has developed its own file management system for short-term archiving, Archer, which was developed to replace the ECS file management system, Unitree. Archer seems to work well under current loads and has the advantage of being tunable to known access patterns at the DAAC.

Locally developed systems, however, can be treated only as temporary solutions because of the continued responsibility to maintain the systems, upgrade them as needed, and migrate them to new platforms. This will drain resources from what should be the primary mission of the DAAC. Whenever possible, the DAAC should use commercial off-the-shelf (COTS) technology. For example, the ECS-provided archiving devices seem reasonable, and the current archiving plans are moving in the right direction with the STK Timberwolf.

Recommendation 4. In order to avoid being distracted from its primary mission, the DAAC should strive to operate in a standardized environment and rely on industry-supported COTS technology whenever possible.

TRMM Support System

The TRMM satellite has two segments—the TRMM mission and two unrelated instruments (the Clouds and the Earth's Radiant Energy System and the Lightning Imaging Sensor). The GSFC DAAC is responsible for archiving and distributing data for the TRMM mission. Originally, the TRMM mission was the responsibility of the Marshall Space Flight Center, and the data were to be processed using the EOSDIS Core System. However, the Marshall Space Flight Center DAAC was closed in 1997, and ECS support of TRMM was first delayed, and then finally canceled. Consequently, the GSFC DAAC had to build the TRMM Support System. The TRMM Support System contains only those functions that its users need; it does not have all the functionality of the ECS. The system was built in eight months with only six full-time equivalents (FTEs), and users appear to be happy with the result. The panel suggests that the GSFC DAAC carefully document its experience with TRMM so that other groups can benefit.

Media Versus Web Distribution Strategy

Last year the DAAC completed WWW-accessible precomputed subsets and accompanying README documentation for all major DAAC data sets. The DAAC plans to (1) continue adding new data products to its anonymous file transfer protocol (ftp) data collections, (2) upgrade the Web-based documentation of existing data sets, and (3) enhance the functionality of the search-and-order Web interface to allow users to do on-the-fly parameter and regional subsetting, to order and track off-line data and documentation, and to use date-specific functions. The panel feels that it would be prudent to focus user interface implementation almost entirely on Web technology.

Connection to the World

The LAN connections between processing and archive computers were designed to handle very heavy traffic (rated at 800 Mb/sec/channel and there are multiple 800-Mb/sec channels). The balance of the system communicates on shared 100-Mb/sec segments. Data distribution to the outside world relies on the NASA Science Internet, which is an OC3 150-Mb/sec line. Although the current networking, both internal and external, appears to be adequate, there are questions as to whether internal networking, at least, is adequate for the loads that will arise with MODIS. The internal rates of communication should be more in balance and higher than currently supported. For the DAAC's internal purposes, 150 Mb/sec is slow. The external rate is probably too low, with many university players in this area moving to OC12 lines.

MANAGEMENT

General Philosophy

The DAAC manager believes that data centers that are driven by requirements are not taking full advantage of the data or serving their customers well. Consequently, the DAAC manager, Paul Chan, has instituted a customer-oriented "business model" for running the center. The business model approach focuses on increasing the demand for use of DAAC data, while decreasing the effort of the customer (shown qualitatively in Figure 3.2). By creating an end-to-end user services group, providing access to data products and services on the Web, simplifying data formats, and creating data products of known or potential interest to a broad community, the DAAC has greatly increased its transaction volume and

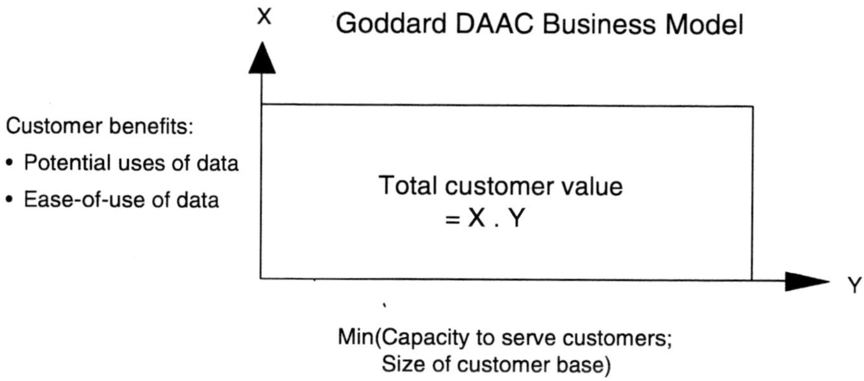

FIGURE 3.2. The GSFC DAAC's business model. SOURCE: Paul Chan, GSFC DAAC.

user base. The panel was impressed with the business model approach and its focus on users.

Implementing the business model required a change in culture at the DAAC. The DAAC manager has overcome the resistance of DAAC staff and the UWG to this approach, and at the time of the site visit, staff morale had improved. Chan delegates responsibility to his staff and they have responded in a positive manner.

Although the DAAC does not have a formal strategic plan document, it has a strong short-term focus on dealing with the AM-1 platform, particularly the large MODIS data sets. Insofar as MODIS is justifiably viewed as the flagship instrument within the AM-1 mission, the panel feels that this focus not only is the correct one, but is in fact critical to the success of the EOS program. The recent delays in launch date have provided a window of opportunity for the DAAC to complete its readiness exercises and to install and test a more flight-ready version of the ECS, thereby mitigating the negative consequences of previous difficulties with the ECS.

Recommendation 5. The DAAC should retain its strong focus on achieving full readiness in time for the AM-1 launch and on being able to secure, archive, and distribute the full MODIS data stream.

Personnel

The GSFC DAAC director has put together an excellent staff. They are professional, motivated, and obviously enthusiastic about their work. They have demonstrated their abilities by successfully assuming responsibility for management of the TRMM data flow. They are open and responsive to criticism and suggestions, which is critically important to their ongoing mission. This respon-

TABLE 3.1. Total GSFC DAAC Costs (million dollars)[a]

	Fiscal Year								
	1994	1995	1996	1997	1998	1999	2000	2001	2002
GSFC DAAC	3.6	3.8	3.4	6.2	6.2	6.0	6.0	6.3	4.7
ECS hardware	0	1.3	6.8	19.2	1.0	7.8	4.4	1.2	1.0
ECS software	0.7	9.5	2.8	1.8	0.4	0.5	0.4	0.3	0.3
ECS personnel	0	0	0.4	0.9	1.5	5.4	6.2	8.4	9.0
Total cost	4.3	14.6	13.4	28.1	9.1	19.7	17.0	16.2	15.0

[a]Budget numbers for FY 1994-1997 are actual values; numbers for FY 1998-2002 are projections, as of May 1998.
SOURCE: ESDIS.

siveness was apparent at the site visit and is also indicated by the implementation of UWG suggestions.

Tension between operations and development is a classic data center problem, particularly for the DAACs because most of the development is being done out-of-house by the ECS contractor. As a result, it is more difficult for the DAAC to shift the emphasis from development to operations as launch approaches. In the past, the DAAC had a poor relationship with the ECS developers, but the arrival of a new ECS liaison, Tom Dopplick, at the DAAC has smoothed tensions between the two organizations. In this capacity, Dopplick works closely and effectively with Chan, and problems are resolved before they reach unmanageable proportions.

Budget

The DAAC's budget grew from approximately $4 million in FY 1994 to $28 million in FY 1997, its peak year (Table 3.1). Although highly variable, the DAAC's average budget is about $15 million per year. ECS hardware, software, and personnel account for nearly 70% of the budget, partly because the DAAC serves as a center for cross-DAAC coordination of the ECS configuration. As a result, the DAAC's hardware and software help support all four AM-1 DAACs.

The DAAC manager takes prides in doing everything as cheaply as possible. For example, the GSFC DAAC acquired the hydrology data from the Marshall DAAC without requiring additional resources, and the DAAC spent only one-fifth of the projected developments costs on TRMM. In fact, until this year the DAAC has always spent less than its approved budget. The DAAC's best measure of cost-effectiveness, however, is that the unit cost has declined from several hundred dollars per order to $60 per order of data.

Contingency Plans

In the event that the ECS is not ready by the AM-1 launch, the instrument teams have been given a choice of which system they will use—Version 0, ECS, a hybrid of the two, or their own home-grown system. In particular, the MODIS instrument team has decided that the MODIS computing facility will be the backup for getting data to the scientists. On the other hand, the DAAC expects that the ECS will be available on time, but without all the promised functionality. If this happens, the DAAC believes it can develop work-arounds to process MODIS data through the ECS. If the contingency plans of the MODIS instrument team are in fact implemented, the DAAC will face a much lighter processing load, may find itself to be oversized, and would have to revise its plans accordingly.

GSFC DAAC AND THE EARTH SCIENCE ENTERPRISE

Relation to Goddard Space Flight Center

The DAAC is hosted by the Goddard Space Flight Center, which provides office and computer space and pays the salaries of the DAAC's civil servants. The DAAC is one of many facilities at the Goddard Space Flight Center and does not receive special recognition from Goddard management. Indeed, since data management is not a central mission of NASA, the DAAC believes its position within Goddard Space Flight Center is vulnerable. The DAAC's primary contact with Goddard management is with Stephen Wharton, chief of the Global Change Data Center, who reports to Vincent Salomonson, director of the Earth Science Division.

Relation to ESDIS

The GSFC DAAC follows the basic roles and responsibilities laid out by ESDIS, but then puts in extra effort serving the customers (the business model). ESDIS neither encourages nor discourages the DAAC from assuming these new responsibilities as long as the basic requirements are met.

At the time of the October 1997 site visit, the DAAC perceived ESDIS as having a development focus and thus a philosophical alignment with the ECS contractor. As a result, tensions with the ECS contractor (see below) led to tensions between the DAAC and ESDIS. Subsequent changes in management at ESDIS and delays in the ECS have created a different kind of problem, which was brought to the attention of the panel in June 1998. There is a growing belief among some in the DAAC that ESDIS as no longer able to enforce standards or interoperability among the DAACs. Instead, EOSDIS is becoming balkanized and it is no longer clear that anyone is in charge of the overall system.

Relation to Other DAACs

In October 1997, the DAAC manager saw the DAACs as being in a friendly competition, although they all followed certain standards and cooperated in some areas, such as driving the direction of the ECS. In addition, the GSFC DAAC worked with the LaRC DAAC on instrument interdependencies and with the EDC DAAC on archive. Even then, however, Chan described the GSFC DAAC as an independent entity within EOSDIS, rather than as a part of the system.

Chan argued that data interoperability (i.e., standard formats that allow use of the same tools on different data sets) is important but that system interoperability, which is burdensome and impedes evolution, should not be an absolute requirement. In the panel's view, this distinction, together with the emergence of a federated system operating over the Web, entails a serious challenge to the requirement for a uniform ECS architecture. This may ultimately call for a profound rethinking and restructuring of EOSDIS in the future.

Relation to the ECS Contractor

The close proximity between the GSFC DAAC and the ECS contractor should, in theory, facilitate a close working relationship between the two organizations. In practice, the DAAC has a poor, but improving, relationship with the ECS contractor. The DAAC has a good relationship with the operations side of the ECS contractor, and the ECS liaison, Dopplick, has greatly eased communication problems between the DAAC and the development side of the ECS contractor by acting as an emissary.

SUMMARY

The GSFC DAAC is a well-equipped, well-run operation, which has a number of impressive accomplishments, such as successfully assuming the management of TRMM data. The DAAC's successful handling of TRMM, SeaWiFS, and other existing data sets bodes well for its ability to handle the large data sets that will result from EOS missions. However, scaling up two orders of magnitude to handle the MODIS data stream is a much greater challenge. MODIS is the flagship instrument of AM-1, and the DAAC's ability to process and disseminate MODIS data and products to users will be a gauge of success for the entire EOS-EOSDIS program.

The success of EOSDIS also depends on the ability of the DAACs and/or other data or service providers to work together to enable users to integrate disparate data sets from a variety of sources. Although the GSFC DAAC has shared processing responsibilities with the EDC, NSIDC, and LaRC DAACs, it views itself as isolated from, rather than an integral partner in, EOSDIS. Stronger links with the other DAACs would help ensure that EOSDIS functions as a system.

Finally, the panel was pleased to find that the DAAC's focus is on users. The DAAC has a number of creative strategies, such as preparing predefined subsets, for meeting the needs of existing users, and it tries to position itself to meet the needs of new user groups as they emerge. The DAAC has also empowered its data teams, which has led to greater user satisfaction with the center. These measures are designed to increase the demand for DAAC data, while decreasing the effort of the users. This is the essence of the DAAC's business model. However, the DAAC's relationship with its primary user community (i.e., scientists) is weak, and the DAAC needs to focus on building relationships with its science teams and the scientific user community, both at Goddard Space Flight Center and elsewhere. Meaningful, ongoing interactions with scientists will help the DAAC understand the impact of its decisions about issues such as reprocessing on the scientific user community.

4

Langley Research Center DAAC

Panel Membership

J.-BERNARD MINSTER, *Chair*, Scripps Institution of Oceanography, La Jolla, California
FERRIS WEBSTER, *Vice Chair*, University of Delaware, Lewes
C. BRUCE BAKER, NOAA National Climatic Data Center, Asheville, North Carolina
ROBERT D. CESS, State University of New York, Stony Brook
RUSSELL R. DICKERSON, University of Maryland, College Park
REX W. TRACY, GDE Systems, Inc., San Diego, California

ABSTRACT

The Langley Research Center (LaRC) DAAC manages data related to atmospheric chemistry and the Earth radiation budget, from remote sensing observations and field experiments. At the time of the review, the DAAC had extended the capabilities of its own information system to manage data from the first Clouds and the Earth's Radiant Energy System (CERES) instrument, which was launched on the Tropical Rainfall Measuring Mission (TRMM) satellite on November 28, 1997. The success of this development and data management effort, which was undertaken because of delays in the development of the EOSDIS Core System (ECS), illustrates the resourcefulness and capability of LaRC DAAC staff. Its success, however, has also led the DAAC to forgo use of the ECS for managing its upcoming data streams. It may be possible for the DAAC to generalize its CERES-specific information system to accommodate other types of data, but to do so would duplicate overall EOSDIS development efforts. It would also incur the risk of isolating the LaRC DAAC from the EOSDIS system, thereby making it more difficult for users to integrate data from the LaRC DAAC with data from other DAACs. Consequently, the panel's main recommendation is that the DAAC develop a transition plan to link its information systems with the ECS and keep a strong focus on the overall EOSDIS goals and ideals.

INTRODUCTION

Langley Research Center began processing Earth Radiation Budget Experiment (ERBE) data in 1985. The LaRC DAAC was formed in 1989 and serves the atmospheric science community, particularly those segments interested in Earth radiation budget, clouds, aerosols, and tropospheric chemistry (Box 4.1). Its current holdings include data from aircraft and satellite instruments and field campaigns.

In addition, the DAAC is currently processing data from the first CERES instrument, which was launched on the Tropical Rainfall Measuring Mission (TRMM) in November 1997. The CERES/TRMM data will increase the volume of the DAAC's holdings from 900 GB to 155 TB by the completion of the mission. Because the ECS was not ready in time for the TRMM launch, the DAAC developed the Langley TRMM Information System (LaTIS) from its existing information management system.

The DAAC is scheduled to receive data from several EOS-era instruments over the next few years. These include two more CERES instruments, the Multi-angle Imaging Spectroradiometer (MISR) and Measurements of Pollution in the Troposphere (MOPITT), all on the AM-1 platform, and the Stratospheric Aerosol and Gas Experiment III (SAGE III) on the Russian Meteor 3M satellite. Incorporation of these data streams will increase the volume of the DAAC's holdings from 3 TB to about 1,300 TB by the completion of the mission, but the DAAC feels that scaling-up is less a data volume problem than a problem with the number and diversity of data sets and the ability to train staff to know enough about the data to help users. If allowed, the DAAC will expand the capacity and functionality of the LaTIS to accommodate these data streams, although this would tend to further isolate the LaRC DAAC from the EOSDIS system.

The Panel to Review the LaRC DAAC held its site visit on November 18-19, 1997. The panel subsequently updated its report through e-mail discussions with the DAAC manager in June through September 1998.

HOLDINGS

The LaRC DAAC has holdings in the areas of the Earth's radiation budget, clouds, aerosols, and tropospheric chemistry (Box 4.2). For example, the long-

BOX 4.1. Vital Statistics of the LaRC DAAC

History. The LaRC DAAC was created in 1989, and since there were no heritage archives, the DAAC could begin with modern technology. The DAAC built its own information management system, which it modified to manage the CERES data.

Host Institution. NASA Langley Research Center in Hampton, Virginia.

Disciplines Served. Atmospheric science; data are available on Earth radiation budget, clouds, aerosols, and tropospheric chemistry.

Mission. To support the Earth Science Enterprise by disseminating information about the Earth System. The scientific priorities of ESE and the U.S. Global Chance Research Program (USGCRP) that the DAAC supports are (1) seasonal-to-interannual climate variability and prediction; (2) decades-to-century climate variability; (3) atmospheric ozone research; and (4) changes in ozone, ultraviolet radiation, and atmospheric chemistry.

Holdings. The DAAC currently holds 3.3 TB of data, and anticipates receiving 155 TB of data from CERES/TRMM and an additional 1160 TB from CERES, MISR, and MOPITT on the AM-1 platform and SAGE III on Meteor 3M.

Users. There were 804 unique users in 1997, of which 30% were outside the United States.

Staff. In FY 1998 the DAAC had 84 staff (including 4 civil servants) and 6 ECS contractors.

Budget. Approximately $7.9 million in FY 1998 (including DAAC costs and ECS-provided hardware, software, and personnel), increasing to $15.3 million in FY 2000.

term radiation budget data from ERBE and the follow-on CERES project will prove valuable to researchers assessing long-term climate variability. An example of an Earth radiation budget data product is given in Figure 4.1. Similarly, the LaRC DAAC enables researchers to view the temporal and spatial variability in the composition of the atmosphere. This provides insight into trends in the oxidizing properties and radiative forcing of the atmosphere. The overview of data points out, for example, areas of photochemical ozone production versus

BOX 4.2. Data Holdings as of January 1998

- *Active Cavity Radiometer Irradiance Monitor II (ACRIM II)*—Total solar irradiance data from October 1991 (ongoing).
- *Earth Radiation Budget Experiment (ERBE)*—Global monthly and daily data from a multisatellite system for November 1984 to an undetermined date.
- *Clouds and the Earth's Radiant Energy System (CERES) Pathfinder*—Data from October 1986.
- *Nimbus-7 ERB*—Global daily data for June 1975 to December 1978.
- *Sulfates/Smoke, Clouds, and Radiation (SCAR)*—Daily data from the eastern United States for July 1993 and Brazil for August to September 1995.
- *Surface Radiation Budget (SRB)*—Global daily and monthly averages for March 1985 to December 1988.
- *Surface Solar Energy (SSE)*—Data for March 1985 to December 1988.
- *International Satellite Cloud Climatology Project (ISCCP)*—Global daily, biweekly, and monthly data for 1983 to 1995.
- *First ISCCP Regional Experiment (FIRE)*—Regional daily data for October and November 1986, June and July 1987, October and November 1991, and June and July 1992.
- *Subsonic aircraft; Contrail & Clouds Effects Special Study (SUCCESS)*—Aircraft- and ground-based measurements for April to May 1996.
- *Aerosol Research Branch (ARB) 48" Light Detection and Ranging (LIDAR)*—Ground station data for January 1982 to an undetermined date.
- *Stratospheric Aerosol and Gas Experiment I, II (SAGE I, II)*—Global monthly data for February 1979 to November 1981 (SAGE I), and global monthly and seasonal data for October 1984 to an undetermined date (SAGE II).
- *Stratospheric Aerosol Measurement II (SAM II)*—Data from the polar regions are available for October 1978 to January 1993.
- *Biomass Burning*—Ground station data for December 1979 to January 1981.
- *Global Tropospheric Experiment (GTE)*—Regional daily data for July and August 1988, July and August 1990, September and October 1992, and September and October 1996.
- *Measurement of Air Pollution from Satellites (MAPS)*—Global daily data for 1984 and 1994 Space Shuttle flights.

BOX 4.2. Continued

- *NASA Water Vapor Project (NVAP)*—Global daily and monthly data for January 1988 to December 1992.
- *Scanning Multichannel Microwave Radiometer (SMMR)*—Global monthly data for 1979 to 1984.
- *Special Sensor Microwave/Imager (SSM/I)*—Global monthly data for July 1987 to December 1991.
- *Atmospheric Radiation Measurement Enhanced Shortwave Experiment (ARESE)*—Regional daily data for September to October 1995.

SOURCE: NASA (1998).

destruction and shows the impact of anthropogenic emissions on the chemistry of the remote troposphere.

Formats

To facilitate integration of different types of data from different sources, EOS data will be distributed in a common format, Hierarchical Data Format (HDF). The LaRC DAAC and its users have little experience with HDF-EOS and are concerned about its being a deterrent to data use and access, particularly due to a limited set of tools for its use. In addition, the three HDF data structures supported by the ECS—point, swath, and grid—do not apply to all LaRC data sets. ESDIS recognizes the problem, but until a solution is devised, the panel counsels the DAAC to consider supporting multiple formats. Several data centers and DAACs have experience writing translators for HDF, and the DAAC should take advantage of this existing expertise before MISR and MOPITT are launched.

Metadata

The scientists involved with the mission or experiment are largely responsible for the metadata associated with their data products. For non-EOS data sets, the LaRC DAAC provides the data producers with a data ingest form, and the scientists specify the key parameters and perform the quality assurance and quality control. The DAAC then checks the data against the acceptable values provided by the scientists as another quality assurance and quality control step for the data. For EOS missions, the DAAC is obligated to use the ECS metadata model, although data providers can also define product-specific metadata. Only a subset of the ECS metadata was used for the CERES/TRMM mission, however,

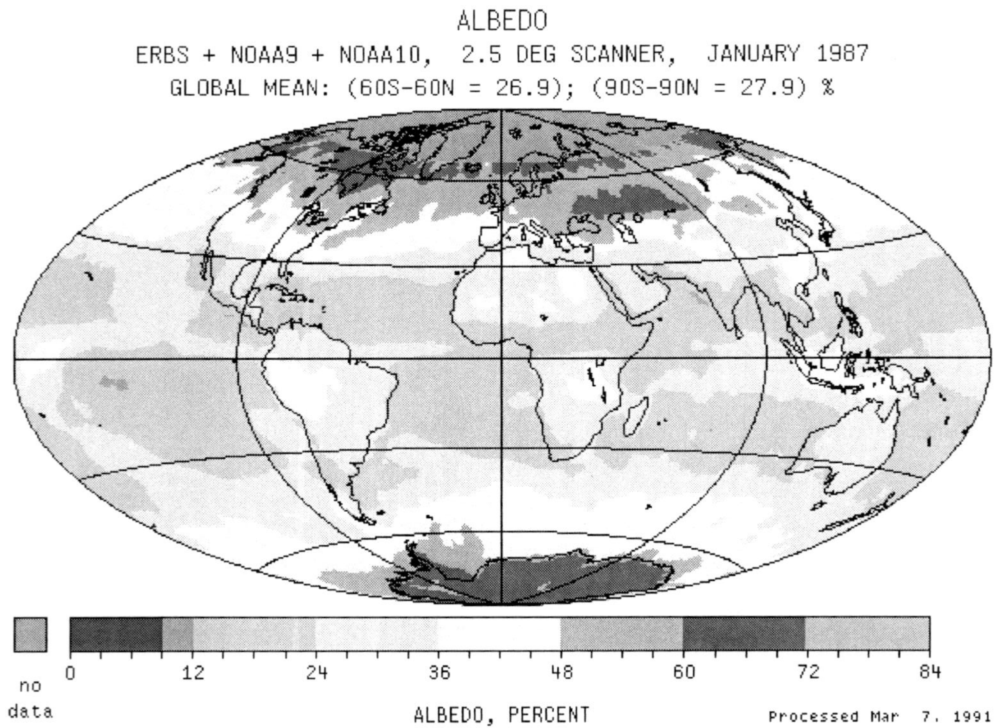

FIGURE 4.1. Global mean albedo for January 1987 based on measurements from the Earth Radiation Budget Experiment (ERBE) scanners on three satellites. With the current data products from the new Cloud and Earth's Radiant Energy System (CERES) instrument on the Tropical Rainfall Measuring Mission (TRMM) satellite, the ERBE and CERES data products will constitute a valuable long-term Earth radiation budget data set. SOURCE: LaRC DAAC.

which may pose difficulties for its incorporation into the ECS. The problem is that the ECS has a minimum set of metadata "valids" without which the ECS will not work.

Processing Plans

At the time of the site visit, the DAAC was preparing for the imminent start of processing for the CERES/TRMM experiment. (The processing began in December 1997, and the DAAC has already reprocessed some of the data.) Similarly, the DAAC's processing plans were to generate Level 1-3 CERES/AM-1, MISR, and MOPITT data, and to distribute and archive the AM-1 data and SAGE III products using the ECS (see Table 1.1 for a description of processing levels). Delays in the ECS, however, have led NASA to consider transferring processing responsibilities for some instruments from the DAACs to the science teams. As a result, the DAAC's backup system, the LaTIS, will be used to process and handle CERES/AM-1 data. MISR processing is still planned for the ECS, and MOPITT processing is planned to transition to the ECS after an initial validation period.

To prepare for the data streams, the DAAC is testing and integrating Project Generation Executables (PGEs) into its information system. As of June 1998, the DAAC had run 31 simultaneous instances of the CERES/TRMM PGE for producing Level-1B data. With regard to MISR, eight PGEs have been delivered to the DAAC for testing, and the Level 3 PGEs are still being developed. The DAAC has successfully run two of the MISR PGEs through the data processing system and is testing and integrating the others into the ECS. Tests to integrate MISR PGEs into the LaTIS are scheduled for summer 1998. Finally, the three MOPITT PGEs will be run by the MOPITT instrument team rather than the DAAC, until six months after launch. Although the DAAC has tested MOPITT PGEs on earlier versions of the ECS, comprehensive tests for integrating the PGEs into the system will not begin until closer to launch.

At the time of the review, the panel was concerned that interdependencies with other instruments could affect the DAAC's readiness for the CERES instruments. CERES/TRMM depends on the Level 1B product of the Visible Infrared Scanner (VIRS), and CERES/AM-1 depends on Level 1B products of the Moderate Resolution Imaging Spectroradiometer (MODIS). Simulated VIRS and MODIS data for testing PGEs are being provided to the LaRC DAAC by the CERES instrument team, and the DAAC anticipates no problems receiving the data it needs for PGE testing.

Reprocessing Strategy

The DAAC does not have a formal strategy for reprocessing. Rather, it attempts to ensure adequate resources to allow both processing and reprocessing, and works with its instrument teams to implement reprocessing as necessary. For example, with the CERES/TRMM experiment, the DAAC has been processing Level-1B data daily since December 1997. Several revisions of the Level-1B PGE have been incorporated, and parts or all of the data have been reprocessed twice as part of algorithm verification. In the longer term, the CERES team and the DAAC are planning a more extensive reprocessing of CERES/TRMM data in the third year of the mission.

Subsetting Strategy

Subsetting is a concern of the DAAC. Subsetting tools are scheduled to be provided in later versions of the ECS, but they will likely not be available before the next EOS mission is launched. Consequently, the DAAC has begun developing capabilities for subsetting large-volume, high-demand products on the fly.

Data Acquisition

The satellite holdings of the DAAC are most useful when combined with other types of data. For example, the LaRC DAAC mandate includes aerosols and tropospheric chemistry. Aerosols and reactive trace gases show a good deal of variability on regional scales; their concentrations are often significantly higher near sources, but certain meteorological conditions can cause pollution on a large scale. These regional-scale perturbations can be relevant to global change issues, for example, the sulfate aerosol and ozone from industrial processes in North America, Europe, and Asia and biomass burning in the tropics. Numerous regionally focused field experiments to gather data on aerosols and tropospheric gas-phase species have been conducted or are planned, and the LaRC DAAC will have to be aggressive about acquiring these data sets in order to best serve the scientific user community.

Long-Term Archive

NASA and NOAA are negotiating a Memorandum of Understanding to transfer data from the DAACs to NOAA archives sometime after the end of the EOS missions. Until this agreement is concluded, the LaRC DAAC plans to continue managing its data as long as funds are available. Although the DAAC sees the data transfer as inevitable, it believes that this will be expensive and that a NOAA archive will move the data too far away from those who have expertise with the data to be really useful.

USERS

Characterization of the User Community

The user community for the LaRC DAAC can be divided into two general types; data providers and data users. The DAAC philosophy is to serve both user types equally, while recognizing that scientific researchers of either type have the highest priority. Data providers are primarily scientists involved with current or recent space experiments, field experiments, data analysis and synthesis teams, and EOS instrument teams. Data users include scientists, educators, the general public, and the commercial sector. DAAC surveys indicate that scientists are the primary user group, accounting for more than 50% of Web-based inquiries, and overseas users are the fastest-growing user group. The DAAC tries to serve its entire user community by providing a variety of ways to access and use its data holdings.

User Working Group

The role of the DAAC's User Working Group (UWG) is to identify new data sets that would be useful to the DAAC and its users, and to determine whether additional functionality is needed. The UWG feels that the DAAC is responsive to its recommendations, but with the completion of Version 0 of the information system, the members feel that their job is done and there is little interest in joining or participating in the meetings. Neither the DAAC nor the UWG feels that the UWG has a role in guiding the development of the LaTIS or ECS systems. After an 18-month hiatus, the UWG held its requisite two meetings in the past year, although only about 25% of members attended.

The panel feels that the UWG is not as effective as it could be. The DAAC seems to want the UWG to become an outreach group that operates under the DAAC. Instead, in the panel's view, the UWG should be an external group that provides scientific oversight, and it should have a co-chair from outside Langley Research Center as mandated in its charter. The panel followed up with the UWG in June 1998 and learned that a co-chair from the University of Colorado had recently been appointed. To ensure that the needs of the users are met the panel urges the UWG co-chairs to become familiar with a broad range of DAAC activities, and to become more aggressive in providing needed advice.

Recommendation 1. To fulfill its responsibilities to the users, the UWG should be assertive and act as an independent body. The DAAC in turn should respond to recommendations of the UWG.

Relationship with the Scientific Community

A significant strength of the LaRC DAAC is its strong interaction with the radiation and clouds scientific community who both produce and use DAAC data. This evolved as a result of Langley's processing of the ERBE data beginning in 1985. Thus, when the DAAC was formed in 1989, the initial staff already had experience in interacting with ERBE scientists, whose close geographical proximity to the DAAC greatly facilitated this interaction. The evolution continued as preparations were made for processing CERES data; the DAAC personnel actively interact with CERES algorithm developers as well as attend CERES Science Team meetings. The proximity and interaction with both ERBE and CERES have clearly proven beneficial to the DAAC staff, providing them with a unique appreciation of the science issues relating to the data they are processing.

Field campaign data are generally more complex than satellite data. They involve one-of-a-kind instruments and require extensive metadata. One such field experiment was the First ISCCP Regional Experiment (FIRE) for which the LaRC DAAC archives and disseminates the data to investigators. LaRC DAAC personnel participated in the planning meetings and field experiments, and scientists from the FIRE experiment felt well served by the DAAC. A representative from FIRE who attended the site visit stated, "The LaRC DAAC is an integral part of our experiment." On the other hand, this strong involvement with specific projects limits the DAAC's ability to take a broad view, and attention should be paid to other instruments and disciplines.

The DAAC's cooperation with the radiation and clouds scientific communities has been excellent. Interaction with the broader scientific community using DAAC data is not as strong. For example, Langley Research Center hosts an outstanding atmospheric chemistry facility employing world-class scientists. They hold seminars and offer scientific expertise to anyone in the DAAC who is interested. Some of the DAAC personnel were drafted from this facility, and to take full advantage of the in-house expertise in chemistry, more such appointments should be made in the future. The panel feels that unless ties with a broader range of scientists at Langley Research Center and elsewhere are strengthened, the DAAC will miss the benefit of a richer source of advice and feedback from its users. For example, scientists actively involved with the DAAC could help identify which complementary data sets to acquire (see "*Data Acquisition*," above).

Recommendation 2. The DAAC should seek daily interaction with a broader range of active researchers through a visiting scientist program and/or through closer ties with atmospheric scientists at the Langley Research Center.

User Services

Users can access the DAAC's information system in a variety of ways, including a graphical interface, a character interface, Web search, and ftp. If a user has a problem with any of the access methods, the DAAC attempts to solve it. If time is required to solve the problem, however, the DAAC will provide access to the data via another mechanism. Problems with media or DAAC-provided software are addressed by DAAC staff. If the problem is with the content of the data file, the problem is forwarded to the data provider. The UWG felt that the DAAC handles unsophisticated customer requests and problems well and that it is also effective in identifying the right data providers for users to talk with.

Subscription services are available at the DAAC, but the system is not automated and bogs down quickly. Future releases of the ECS will have subscription services, and the DAAC plans to wait for this capability to be offered by the ECS, rather than develop its own automated service.

Documenting Scientific Productivity

The DAAC will ultimately be evaluated on the number and impact of reviewed scientific publications that acknowledge DAAC services. Although the DAAC has played an integral role in major experiments such as ERBE and FIRE, little effort has been made to track publications or citations to document the productivity of the DAAC. The DAAC presented a list of 12 peer-reviewed journal articles that used DAAC data. Based on the panel's knowledge of the literature, the publications of panel members, and the number of researchers using the DAAC (several hundred per year), the panel suspects that the number of articles that actually used data from the DAAC is probably an order of magnitude greater. DAAC personnel expressed reluctance to make the effort to document their impact on the science, but doing so could help the DAAC avoid unfair criticism in the future. Active, atmospheric researchers on-site (see recommendation above) would have a vested interest in enhancing and documenting the scientific productivity of the DAAC.

Foreign Access

The LaRC DAAC is located within a secure facility, but because few people need to physically visit the DAAC, high security was not viewed by the DAAC as a barrier to access. Visits can be arranged easily for U.S. citizens and residents with green cards. For other individuals, a visit of less than two weeks can be approved by the deputy center director with about ten days notice, and longer visits require approval by NASA Headquarters and about one month notice.

TECHNOLOGY

The Version 0 System

The Version 0 system runs on a collection of networked computers under the UNIX operating system. This system, which is being phased out, offers a standard X-Windows interface, a character-based interface, and an evolving Web-centric interface, all developed at the DAAC. In addition, Version 0 includes a cross-DAAC interoperable Web gateway interface. Some users have written single-purpose software programs that rely on Version 0. The DAAC will have to provide these users with an easy path to upgrade to the ECS baseline in order to be able to quickly retire the Version 0 hardware, software, and processes once an operational version of ECS has been delivered. Only by explicitly planning this transition in a timely fashion will the DAAC be able to eliminate the need to allocate resources to the Version 0 system.

The EOSDIS Core System

The future hardware, software, and process suite specified in the ECS contract comprise the ECS baseline. At the time of the site visit, some hardware with initial checkout and installation software had been delivered, and the DAAC was awaiting the remainder of the baseline system without knowing exactly what the system would include or how it would operate.

Hardware. The hardware associated with the ECS baseline is a modern system architecture composed of SGI Challenge and Sun Ultra Enterprise machines, along with fairly large tape archive systems (EMASS AML and STK PowderHorn) built by Storage Technology. This hardware appears to be fully installed and available for use once the software is delivered. During the site visit the panel was given no evidence that the system hardware had been sized to fit the load expected at this DAAC. Since the aggregate input/output and processing bandwidth available within the installed equipment seems to support the T3 communications link, one must assume that this analysis has been done as part of determining the hardware baseline by the ECS contractor.

The panel notes that the DAAC is accumulating maintenance costs on ECS hardware that sits idle until the ECS software is delivered. If and when the software is delivered, the DAAC will have to make a concerted effort to migrate its data sets to the new system.

Software. Although details of the ECS software design and selection were not available to the panel, the software is described as an object-oriented architecture, composed of approximately 70 COTS packages surrounded by more than a million lines of contractor-developed custom code. This COTS software appears to have grown into a fairly stable set of supported elements as the vendors' products have matured and the ECS contractor has firmed up its design.

Processes. The descriptions given to the panel were unclear concerning time lines and processes necessary to install a locally desired change into the delivered software. LaRC DAAC personnel had only a minimal amount of system knowledge in this area, and it seems likely that staff will need process training when the system baseline is brought on-line later this year. DAAC staff were unsure of what metrics will be available to allow them to understand how well the system is running when it is on-line. Again, basic ECS system training, with an emphasis on production metrics relevant to the LaRC DAAC, would significantly improve the DAAC staff's understanding of the system.

The Langley TRMM Information System

Because the ECS baseline software was not available in time to support the launch and initial operation of CERES/TRMM, ESDIS asked the LaRC DAAC to develop a parallel system that would serve as an interim baseline until the ECS was installed. The resulting system, the LaTIS, was designed to use the COTS chosen by the ECS contractor where possible, based on the rationale that the training requirements for the DAAC staff would be minimized.

Hardware. The hardware baseline for the LaTIS is based on the next-generation SGI machines (the Origin series), which are largely compatible with the ECS Challenge series machines. Both are multiprocessor architectures and use the same processors. However, the processors are connected in different ways and thus require different version of the SGI operating system. The DAAC's experience is that applications generally run on either series, but drivers do not. It expects that the LaTIS would run on Challenge machines, but has not tried this.

LaRC DAAC management obtained a factory-refurbished tape archive (the STK PowderHorn library with six Redwood SD-3 drives) to save money. The archive is completely compatible with the ECS baseline and can therefore become part of the eventual ECS system.

Software. Because the LaTIS contractor is not the same as the ECS contractor and because details of the ECS software design were not available, the LaTIS software was implemented using software that is at least partially incompatible with the current ECS baseline. Specifically, the LaRC DAAC team has chosen to use the BigSur database technology on top of the Informix Universal Server to implement the LaTIS. BigSur is the result of work at the University of California, Berkeley, and has been modified to handle the load and requirements for automating the CERES production process. Although this is a technologically sound choice for the LaTIS design, such incompatibilities will force the LaRC DAAC to operate a separate maintenance process until such time as this software can be merged into the ECS. In other software selections, the LaTIS appears to have achieved compatibility with ECS by using libraries obtained directly from the ECS contractor. When asked about the configuration management of these libraries, however, DAAC personnel indicated that the libraries used for the LaTIS may not have kept up with the latest changes to the ECS libraries. Thus, it appears that this is a second area in which the LaTIS may prove challenging to merge with the ECS.

Langley Research Center scientists have identified a number of additional capabilities, such as subsetting, that could be included in the LaTIS (and the ECS, if they are not already part of that baseline). It is obvious that, if allowed, the LaRC DAAC will expand the functionality of the LaTIS to meet the needs of its users. Thus, it becomes even more important to ensure that the system is covered by a technically sound set of change and upgrade processes if and when the ECS can be used.

Processes. The processes used for the LaTIS appear to be reasonably well thought out and consistent with modern engineering procedures. The DAAC uses a configuration management mechanism for identifying, implementing, testing, and installing changes to the baseline, as well as a defined end-to-end test procedure using both live instruments and simulated data. In addition, the LaTIS hardware configuration includes a smaller, similar system in parallel to the main processing system where additional development and/or changes can be tested prior to cutting in a new production baseline. Thus, the LaTIS appears to be a technologically sound production system. However, the DAAC does not keep metrics for measuring its performance, making progress difficult to document.

Recommendation 3. To monitor the performance of the LaTIS, the DAAC should document its production metrics.

The LaTIS presents an enigma to the DAAC. Although the system looks as though it works, it requires an infrastructure and budget for its lifetime support. Such an infrastructure will quickly be recognized as redundant with the ECS baseline once that system is installed and operational (see Recommendation 4 below).

Media Versus Web Distribution Strategy

The DAAC is experienced in the use of conventional methods of providing data to its customers. Such methods include responding to user requests by providing data on magnetic tape, floppy disk, or CD-ROM. A majority of users are now requesting that data be provided via the Internet. Statistics on users' access methods show that in nine months (December 1996 to September 1997), 70% of users have switched from ordering data off-line to ordering data over the Internet. To serve these users, the DAAC has made good progress on developing its Web site; the panel found the Web site to be simple, straightforward, and well documented, although it can be difficult at times to find specific data sets.

MANAGEMENT

General Philosophy

Since the time of the review, the DAAC manager, Roy Dunkum, had retired and Richard McGinnis, the deputy, has become DAAC manager. Although the DAAC had only nine months to prepare for CERES/TRMM, and delays in the ECS had compressed the time in which to prepare for the AM-1 platform, Dunkum seemed confident about the DAAC's ability to handle the upcoming data streams. This assessment seems to be based on the DAAC's successes in managing existing data sets and in developing the Version 0 and LaTIS systems, but whether the DAAC can manage the "fire hose" of data remains to be seen.

Dunkum's philosophy was that for system design, hardware is cheap and it is reasonable to buy sufficient hardware to accommodate average peak loads. For staffing purposes, a philosophy of "just-in-time" hiring has been followed because hiring staff before there is work to do wastes money and results in low morale. Just-in-time staffing, however, may make it difficult for the DAAC to maintain the high level of training that it has provided to staff in the past.

Personnel

The panel perceived serious tensions between DAAC staff and the ECS contractors on-site. The ECS contractors were not invited to help prepare for the NRC site visit and they were not introduced to the panel. Uncertainty about the ECS contract is high, and the ECS contractors are worried about job security. Emergency backup plans such as the LaTIS, which work around the ECS, and proposals to replace the ECS with the LaTIS have made matters worse. Even with these morale problems, however, staff turnover (five or six individuals per year) is probably normal.

Budget

The LaRC DAAC's total budget is $7.9 million in FY 1998, with an average budget over a nine-year period of $10.8 million (Table 4.1). The main variability in the budget is related to major hardware acquisitions (e.g., FY 1996 and FY 1997 for CERES/TRMM).

To show its cost-effectiveness, the DAAC cited several factors, including just-in-time staffing; the purchase of a used and upgraded StorageTek; the rela-

TABLE 4.1. Total LaRC DAAC Costs (million dollars)[a]

	Fiscal Year								
	1994	1995	1996	1997	1998	1999	2000	2001	2002
LaRC DAAC	1.6	1.7	2.6	8.6	7.2	8.6	11.0	13.3	10.3
ECS hardware	0	0.6	6.5	6.7	0.3	7.6	3.5	1.2	0.4
ECS software	0	0.1	0.8	0.4	0.2	0.4	0.2	0.2	0.2
ECS personnel	0	0	0.1	0.1	0.2	0.7	0.6	0.6	0.6
Total cost	1.6	2.4	10.0	15.8	7.9	17.3	15.3	15.3	11.5

[a]Budget numbers for FY 1994-1997 are actual values; numbers for FY 1998-2002 are projections, as of May 1998.
SOURCE: ESDIS.

tively low cost of living; the large pool of expertise in the area, which keeps relocation expenses low; and low growth in the budget, compared with the growth in data sets, data volume, projects, and granules. Quantitative measures of cost-effectiveness are not kept by the DAAC.

Contingency Plans

The DAAC has two types of contingency plans. The first, development of the LaTIS to manage CERES/TRMM data, is described above. The second is related to the "25-50-75 scenario" for processing data from the AM-1 instruments. In the event of further delays in the delivery of the ECS, NASA has decided that only 25% of the data will be processed initially. Unfortunately, interpretations of what 25% means differ between the DAAC, the ECS contractor, and the instrument teams. The ECS contractors plans to install only 25% of the hardware, but the CERES instrument team plans to process a full month of data at once every four months, which will require full processing power in order to keep up with the data streams. The panel suggests that the interested groups work together to devise a compatible implementation of the 25-50-75 scenario.

Strategic Plans

As noted above, at the time of the site visit, the DAAC had made no real preparations for the ECS. When the ECS is delivered, the DAAC will have to devote a great deal of time, resources, and development to make the systems compatible or to transition its data sets to the new system. Although the rapid development effort that led to the LaTIS is commendable, the panel is concerned about the ability of the current instrument-specific version of the LaTIS to fit within a multi-DAAC system. On the other hand, parallel operation of the LaTIS and ECS systems is not likely to be cost-effective.

Recommendation 4. The DAAC should prepare a strategic plan either to achieve full functional compatibility between the LaTIS and the ECS or to effect a transition from the LaTIS to the ECS, whichever is the most cost-effective way to maintain compatibility with the remainder of the DAAC system.

LARC DAAC AND THE EARTH SCIENCE ENTERPRISE

Relation to Langley Research Center

The DAAC is separated physically and programmatically from the atmospheric scientists at Langley Research Center. The DAAC is located in the Information Systems and Services Division of the Internal Operations Group of NASA's Langley Research Center. The Atmospheric Sciences Division, on the other hand, is located within the Space and Atmospheric Science Program Group. The heads of both groups report to the Langley Research Center director.

As indicated by their presence at the site visit, Langley managers are interested in the operations of the LaRC DAAC. Langley Research Center provides facilities, secretarial support, and travel support to the civil servants. In the future, the DAAC will begin "full-cost accounting" and will have to pay for all the facilities and services it uses.

Relation to ESDIS

The LaRC DAAC identified two problems in its relationship with ESDIS—communications and philosophical differences. With regard to the former, UWG members complained that action items on software sent to the ESDIS Project are mostly ignored. However, the primary communication problem—that ESDIS acts as a gatekeeper for interchanges between the DAAC and the ECS contractor—has been resolved. The DAAC now discusses bugs, software fixes, and so forth, directly with the ECS contractor.

Philosophically, the DAAC does not believe that a uniform EOSDIS, which is promulgated by ESDIS, is in the best interests of its users. It therefore works to customize EOSDIS functions and to take on some of the responsibilities of the ECS contractor. For example, to use people and resources efficiently, Dunkum received permission and funding from ESDIS to provide maintenance on the ECS hardware and software already at the DAAC. The development of the emergency system LaTIS, which was built to fewer requirements than the ECS, is another example. If allowed, the DAAC would likely attempt to extend the capabilities of the LaTIS to meet the needs of AM-1 instruments, rather than use the ECS.

Relation to Other DAACs

In the past, all of the DAACs cooperated to create interoperability in Version 0 and the ECS, but now that system development has become the responsibility of the ECS contractor, communication between the DAACs has largely been reduced to regular teleconferences. The LaRC DAAC works with the GSFC DAAC because of instrument interdependencies, but is not proactive in marketing its own tools (i.e., the LaTIS) or learning about the tools or techniques of other DAACs. In the panel's view, these examples illustrate a weakness in the links between the LaRC DAAC and the other DAACs. The DAAC's anticipated use of the LaTIS, rather than the ECS, for upcoming EOS data streams will likely further isolate the LaRC DAAC from the system.

Relation to the ECS Contractor

Although the DAAC finds dealing with the ECS contractor in Landover to be slow and difficult, it feels that the ECS contractor is becoming more responsive to the identification of bugs. Nevertheless, the DAAC does not believe that the ECS, if it is ever delivered, will meet the needs of the science teams. Not only would the ECS have to be adapted to take advantage of the new network environment (e.g., WWW, Java), but new software demanded by the science teams would also have to be incorporated into the ECS release schedule.

Despite these problems, the panel found the DAAC to be complaisant about the ECS. Indeed, with the diversion of ECS maintenance funds to the LaRC DAAC, and ECS engineers to the development of the LaTIS, the LaRC DAAC feels it has already worked around the anticipated ECS difficulties.

SUMMARY

Important functions of DAACs are to make it easy for data providers to work with the information system and to satisfy their users. The LaRC DAAC's interactions with its data providers are superb. The DAAC even participates in the design and execution of data management aspects of the field experiments, which will help it manage the resulting data better in the long term. The DAAC's relationship with users, however, is not as strong. In particular, the DAAC could benefit from having more interaction with the atmospheric science user community. A ready opportunity to understand more about the DAAC's data sets exists through a closer relationship with the atmospheric scientists who are housed just a few buildings away from the DAAC. In addition, the User Working Group, which is seeking a role in the EOS era, could help the DAAC learn more about user needs, as well as the utility of the DAAC's formats and tools.

Given the delays in the ECS, the development of tools to make the DAAC's data easier to find and work with will be increasingly important to the user community. The DAAC manages its existing data sets well, but to make EOS data as useful, the DAAC must continue developing its Web interface and subsetting tools. The latter will be particularly important for dealing with large EOS data sets.

The DAAC has already begun to deal with its first EOS data stream from the CERES instrument, which was launched on the TRMM satellite in November 1997. Because the ECS was not ready, the DAAC developed a CERES-specific information system, the LaTIS, to manage the data. That the LaTIS appears to be working well is a tribute to the talent, dedication, and resourcefulness of DAAC staff. The use of the LaTIS, however, comes with two costs. First, the DAAC incurs maintenance costs for the information system for the lifetime of the mission. Second, the LaTIS was built to fewer specifications and a smaller metadata model than the ECS, which may pose downstream difficulties for users trying to obtain data from the different information systems. As such, the LaRC DAAC is in danger of becoming a weak link in the ultimate EOSDIS system.

5

EROS Data Center DAAC

Panel Membership

J.-BERNARD MINSTER, *Chair*, Scripps Institution of Oceanography, La Jolla, California
FERRIS WEBSTER, *Vice Chair*, University of Delaware, Lewes
RUTH S. DEFRIES, University of Maryland, College Park
ALLEN M. HITTELMAN, NOAA National Geophysical Data Center, Boulder, Colorado
DAVID A. LANDGREBE, Purdue University, West Lafayette, Indiana
DAR ROBERTS, University of California, Santa Barbara

ABSTRACT

The Earth Resources Observation Systems (EROS) Data Center (EDC) DAAC is hosted by the U.S. Geological Survey. Its mission is to manage and distribute data products generated by low-Earth-orbit missions of the EOS program that deal with land surfaces and land processes. At the time of the formal review, the DAAC had no experience with the EOSDIS Core System (ECS), which is intended to provide the distribution and subsetting capabilities for all of the DAAC's new data streams. The DAAC has since worked with the ECS contractor on its "mini-DAAC," and launch dates have been delayed by at least six months. The extra time should enable the DAAC to address, at least partially, the panel's main recommendations: (1) the DAAC should devote considerable effort to preparing for the Landsat 7 and AM-1 data streams, including developing contingency plans for inevitable initial difficulties with the ECS; and (2) the DAAC should develop a clearer vision for serving the needs of its scientific and, potentially much more numerous, applications users. This vision should include a plan for meeting the challenges of the EOS era, including major increases in data volume and product diversity, a broader user community, and concomitant increased complexity of user services.

INTRODUCTION

The EDC DAAC was created in 1992 to manage data from Landsat 7 and other land remote sensing instruments. It is located within the EROS Data Center, which has been managing land processes data for two decades, but its holdings and operations are separate from those of the USGS. The creation of the EDC DAAC signaled a shift in the focus of the Landsat program. Previous Landsat satellites served primarily the applications community, a very large user group that is likely to increase substantially if data become available at lower cost and in near real time. Landsat 7, on the other hand, has been incorporated into the EOS program as a science instrument, and scientists will be the DAAC's highest-priority users (Box 5.1). However, the demands on the system will likely still be driven by the applications users.

When it was formed, the EDC DAAC did not acquire any heritage data sets. Its data sets, which are few in number but relatively large in size, include the Global 1-Kilometer Advanced Very High-Resolution Radiometer (AVHRR), Landsat Pathfinder, and the digital elevation model Global 30 Arc-Second Elevation Data Set (GTOPO30).

In the EOS AM-1 era, the DAAC will manage data from Landsat 7 and the Advanced Spaceborne Thermal Emission and Reflection Radiometer (ASTER). It will also receive Moderate-Resolution Imaging Spectroradiometer (MODIS) data from the GSFC DAAC and will distribute and possibly process MODIS land products. Data from these instruments, which are scheduled to be launched in 1999, will increase the volume of holdings by a factor of 20, from 9 TB to nearly 200 TB per year (Box 5.1).

To prepare for these data streams, the DAAC has requested an additional 50 ECS contractors (there are 38 ECS contractors at the DAAC in FY 1998), and it has begun readiness exercises. The DAAC has tested the algorithms for processing ASTER data and appears to be ready for the ASTER data stream. This is to be credited in large part to the ASTER instrument team, which was extremely diligent in porting its processing codes to the DAAC and testing them extensively. Readiness for MODIS will depend, in part, on the GSFC DAAC and the MODIS science team; at the time of the review, the schedule and products from the MODIS science team were still uncertain, and only limited tests of the algorithms had been completed.

Delays in the delivery of the ECS and uncertainties about its capabilities

BOX 5.1. Vital Statistics of the EDC DAAC

History. The EDC DAAC was created in 1992. Its roots go back to 1972 when the long-standing, successful relationship between NASA and the USGS to archive, process, and distribute Landsat data was established.

Host Institution. USGS EROS Data Center in Sioux Falls, South Dakota.

Disciplines Served. Biology, hydrology, limnology, and ecology; data are available on processes existing and operating at or near the land surface.

Mission. Archive, process, and distribute EOS land processes data for use by the earth science and global change research communities.

Holdings. The DAAC currently holds 9 TB of data and anticipates receiving more than 50 TB of data per year from Landsat 7 and an additional 80-145 TB per year from the AM-1 platform.

Users. There were 1156 unique users in FY 1997.

Staff. In FY 1998 the DAAC had 4 civil servants, 68 FTEs, and 38 ECS contractors.

Budget. Approximately $6.5 million in FY 1998 (including DAAC costs and ECS-provided hardware, software, and personnel), increasing to $13.7 million in FY 2000.

have made preparations for the EOS data streams more difficult. The EDC DAAC relies exclusively on the ECS for managing the data, and a relatively high level of functionality will be required to ensure timely processing, subsetting, and distribution of data.

The Panel to Review the EDC DAAC held its site visit on November 24-25, 1997. The following report is based on the results of the site visit and e-mail discussions with DAAC personnel held in June and July 1998.

HOLDINGS

The data currently distributed by the EDC DAAC (Box 5.2) are widely used for a variety of scientific applications. The Landsat Pathfinder data, for example, serve as a rich resource for many aspects of global change studies related to the

BOX 5.2. Data Holdings as of January 1998

- *Global 1-Kilometer Advanced Very High-Resolution Radiometer (AVHRR)*—Global and regional 1.1-km resolution data beginning in 1992.
- *Spaceborne Imaging Radar-C (SIR-C)*—Six-hour global data from 1994 Space Shuttle flight.
- *Global 30 Arc-Second Elevation Data Set (GTOPO30)*—30 by 30 arc-second digital elevation data.
- *Landsat Pathfinder*—Seasonal global and regional data from the 1970s, 1980s, and 1990s.
- *Aircraft Scanners*—Monthly local data beginning in 1987.

SOURCE: NASA (1998).

land surface. The 1-km AVHRR data have been used by many researchers to characterize the land surface as well as to model the role of the land surface in earth system processes. GTOPO30 topographic data (Figure 5.1) are needed for both radiometric and geometric correction of sensor data. Use of these data and imagery has greatly facilitated research on topics ranging from land cover change to geomorphology to hydrology.

Landsat 7

Three agencies are involved in Landsat 7—NASA, NOAA, and USGS. NASA is responsible for building and launching the spacecraft and instrument, and for building, installing, and testing flight operations, ground data reception, and processing, archive, and distribution systems. It is also funding the DAAC to provide Level 0 archive and distribution for NOAA and Level 1 product generation and distribution for NASA (see Table 1.1 for a description of processing levels). NOAA is responsible for coordinating the international ground station network and for performing flight operations, data acquisition, and preprocessing. Finally, the USGS is responsible for long-term archive and distribution. Such sharing of responsibility is a concern to the panel because no single entity has end-to-end responsibility for the program.

Formats

Nearly all EOS data sets will be formatted in Hierarchical Data Format (HDF). The land processes community has little experience with HDF-EOS, but

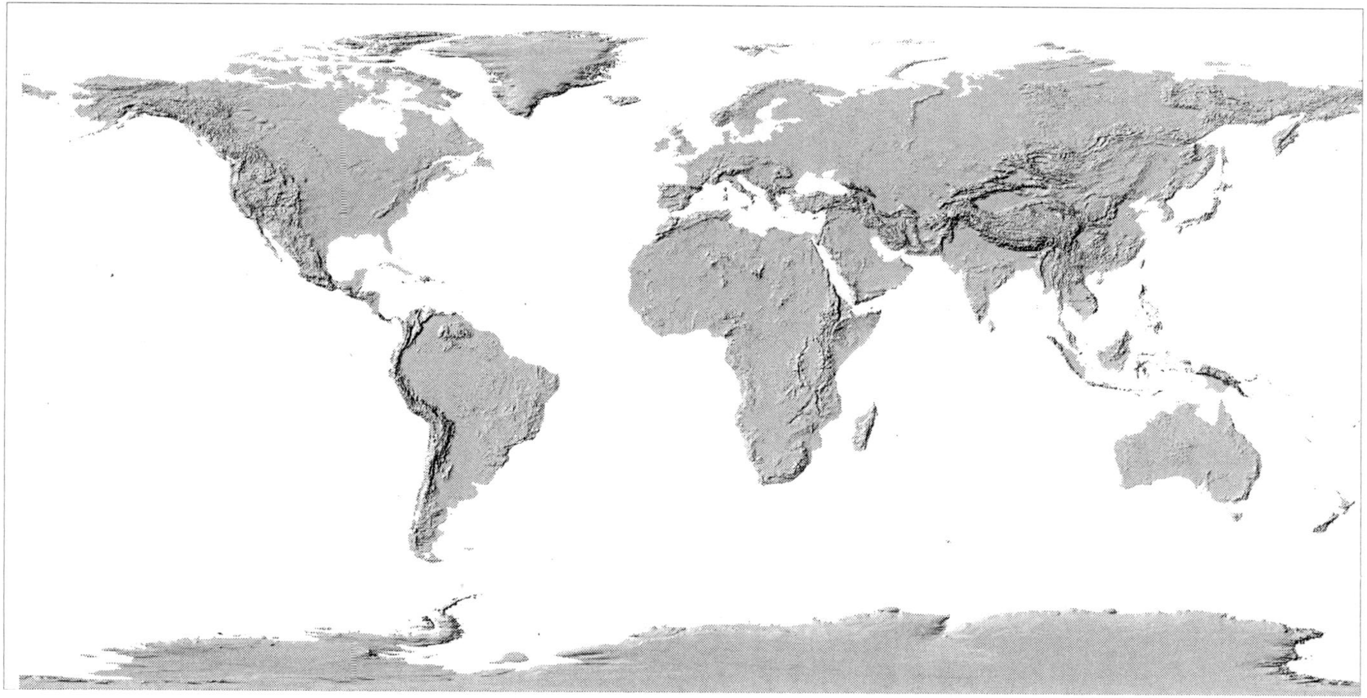

FIGURE 5.1. Global shaded relief map prepared from GTOPO30, a global digital elevation model with elevation values regularly spaced every 30 arc-seconds on a latitude and longitude grid, which equates to a spacing of approximately 1 km. GTOPO30 was developed at the EDC DAAC to meet the topographic data requirements of products from other EOS instruments. SOURCE: EDC DAAC.

the DAAC could undertake efforts immediately, such as converting existing data sets to HDF-EOS for data distribution and exchange.

Data Restrictions

Currently all DAAC data are unrestricted, but the EDC DAAC is planning to buy and disseminate commercial Landsat data from foreign ground stations. Use of these commercial data will be subject to restrictions established by the commercial vendor, the Earth Observation Satellite Company.

Processing Plans

At the time of the site visit, the EDC DAAC was scheduled to receive and process data from three remote-sensing instruments—Landsat 7, ASTER, and MODIS. Delays in the ECS, however, have led NASA to consider transferring data processing from the DAACs to the science teams on an instrument-by-instrument basis. If the DAAC and instrument teams adhere to the original processing plan, the DAAC will receive, archive, and disseminate Level 0 Landsat data for NOAA and will later produce and distribute Level 1 Landsat products for NASA. With regard to ASTER, the DAAC will receive Level 1 data from Japan and will archive and distribute it for NASA. The DAAC will also produce and distribute Level 2 products on demand on behalf of the U.S. ASTER Science Team. Finally, the DAAC will receive Level 2 MODIS data from the GSFC DAAC and will process, archive, and distribute Level 2+ MODIS land products on behalf of the MODIS science team.

To prepare for these data streams, the DAAC is testing Product Generation Executables (PGEs) for ASTER and MODIS, and integrating them into the ECS. The DAAC has received all the PGEs for ASTER and has integrated them as far as is possible in the current release of the ECS. Since the full, required capabilities of the ECS have not been delivered to the DAAC, the PGE-integration process is not yet complete. With regard to MODIS, only two of 33 PGEs have been delivered to the DAAC, and neither has been integrated with the ECS. In this case, the readiness of the EDC DAAC to produce MODIS land products is limited mainly by the readiness of the MODIS science teams and GSFC DAAC.

On the other hand, if some or all of the Landsat, ASTER, and MODIS instrument teams decide to process the data themselves, the DAAC's role will be limited to traditional data management tasks (e.g., dissemination, archive, user services; see Chapter 2, "*DAAC versus Data Center*").

Reprocessing Strategy

Reprocessing Level 1 and higher data to accommodate improvements in calibration and algorithms is necessary to keep data sets viable and internally consistent. Anticipating that the instrument teams will want data to be reprocessed often, the DAAC plans to work with its Science Advisory Panel to set priorities on reprocessing.

Subsetting Strategy

As noted above, data sets held by the EDC DAAC are large and will have to be subsetted to make them manageable to the user community. Rather than developing its own subsetting tools, the EDC DAAC plans to use the subsetting capability of the ECS. ECS subsetting, however, will not be available until mid-1999 at the earliest, long after launch of the AM-1 platform. The impact of the DAAC's decision on users is described below (see "*Data Set Preparation*").

Long-Term Archive

NASA and USGS have signed a Memorandum of Understanding stating that DAAC holdings will move to the USGS three years after the conclusion of the mission(s). The USGS has a line item in its budget to pay for the transition. The mission of the EROS Data Center includes long-term data storage, and the institution has successfully carried out this mission since its inception in the 1970s. Given the importance of maintaining the archive for global change science, the panel suggests that the DAAC begin to plan, at an early stage, for the transfer of data from the DAAC to the EROS Data Center for long-term archiving.

USERS

Characterization of the User Community

Much of the current and expected user community is now assumed to be drawn from the Earth science research community, and it is apparently assumed that the instrument teams for the EOS instruments are representative of this community. However, given the greater breadth and variety of the EOS instrument suite, the user community after the launch of the AM-1 platform and Landsat 7 is likely to be much larger and more diverse than has been the case for previous, less complex instrument sets. It is of critical importance that significant effort be made to predict the size, character, and makeup of this larger, more diverse user community and to assess its needs. With such a user profile, it should be possible to construct a data archive and access system that optimally fits the client community.

For example, the Landsat series of satellites has a unique history, one unlike other instruments in the EOS program. Early in the life of the Landsat program, NASA placed the emphasis of this program on applications rather than science *per se*. Much of the user community for Landsat is therefore composed of what might be referred to as remote sensing practitioners (e.g., value-added commercial organizations; nonscientific federal, state, and local government officials; educators; the general public).

Throughout the 25 years of the Landsat program, substantial emphasis has been placed on "data continuity" because of these users. It was not until the incorporation of Landsat 7 into EOS that Landsat came to be regarded primarily as a science device, and to be treated in this fashion. Given that the needs and expectations of the science and nonscience collection of users are quite different, both must be taken into account in the design and operation of the EDC DAAC, if it is to be successful. Parenthetically, it seems possible that this broadening of the user community will extend to other EOS instruments as the data become available.

Recommendation 1. Based on a broad range of inputs, the DAAC should construct a profile of its expected user community for AM-1 and Landsat 7 data sets. This profile can serve as a sound basis for optimizing the active archive and its user interface to meet the needs of a broader community, and for fostering a sense of ownership among the users.

Science Advisory Panel

The Science Advisory Panel (SAP) is a formal, government-appointed body. Its current focus is on resource allocation, outreach, and user services for the AM platform, and it also helps the DAAC set priorities on holdings and reprocessing requests. Members feel that they have a good relationship with the DAAC and that the DAAC is fairly responsive to their recommendations.

The SAP is composed of instrument team scientists, researchers, including non-EOS researchers, and a user from the private sector. As noted above, however, the potential user community is much broader (including, for example, "remote sensing professionals") and is likely to expand further in the future.

Recommendation 2. The DAAC should review and adjust the membership of its advisory panel to obtain a balanced view from both scientific and nonscientific users.

Relationship with the Scientific Community

The long-term health of the DAAC depends on a symbiotic relationship between the data operations activities and scientific applications. Data operations could benefit greatly from the participation of scientific users in the design of data delivery systems. Conversely, scientific users would benefit from understanding the constraints in the operations. The location of the DAAC within the

EROS Data Center creates an opportunity for such interactions. There is a rich resource of local scientists who are currently using data provided by the DAAC in their research. These scientists could provide a strong scientific backbone for DAAC activities.

However, the panel observed that interaction between DAAC personnel and the science community has been limited to date. The Science Advisory Panel serves as the primary linkage between the science community and the DAAC. Participation of DAAC personnel in scientific research is not actively encouraged, and there are no positions established for working scientists. Some limited efforts have been made to elicit feedback from scientists in the EROS Data Center on the design of Version 0, but these efforts have not received funding or priority. The observation that the DAAC's operations could be strengthened by involvement from scientific users was confirmed by DAAC managers. They expressed a desire to develop mechanisms that would foster interaction between DAAC personnel and the scientific users. In fact, the DAAC scientist distributed a draft "white paper" containing suggestions to expand the scientific involvement.

Examples of mechanisms that might foster the desired symbiotic relationship between the data operations and the scientific operations include the following:

- a visiting scientists program for research using DAAC data sets to be carried out in-house—the visiting scientists would then provide useful feedback on the design of the data delivery systems;
- involvement of DAAC personnel in research activities, particularly those being conducted within the EROS Data Center—this could be achieved through a short leave of absence from normal DAAC responsibilities or an allocation of a small percentage of the staff member's time to a research project;
- establishment of a small number of positions within the DAAC for scientists conducting research using DAAC products—the scientists would provide the perspective of the users in the design of DAAC operations and would be expected to be involved in such activities; and
- encouragement of day-to-day interaction between local scientists and DAAC personnel—this might be achieved through informal users' groups that would provide feedback to the DAAC or other mechanisms that promote informal interactions.

In particular, the panel encourages the DAAC to foster scientific connections with local EROS Data Center scientists who represent a pool of highly qualified users and could provide feedback regarding user services and scientific applications.

Recommendation 3. The EDC DAAC should develop and implement measures to foster working-level interactions between DAAC

personnel and researchers. These measures might include a visiting scientist program, active DAAC participation in scientific research projects, and day-to-day interaction between local scientists and DAAC personnel.

User Services

The successful construction and operation of the DAAC, as for any other service organization, is critically dependent upon (1) knowing to whom the service is to be provided and (2) knowing specifically what the goals of the service are to be. It is important to be as quantitative and specific as possible in answering both questions.

At scientific data centers, it is important that the user services group be able to answer questions of a scientific and technical nature. Thus, a scientific background for user services staff is highly desirable. At the EDC DAAC, the user services group appeared to have trouble answering questions at an impromptu demonstration for the panel, an impression that was confirmed by the SAP. The SAP also pointed out that the DAAC has no system for tracking user requests or the DAAC's responses, thus making it more difficult for the DAAC to serve its customers.

DAAC managers indicated that they plan to hire scientists to be part of the users services team. The panel concurs that improved technical support is needed and recommends that this plan be implemented as soon as possible.

Recommendation 4. The DAAC should implement its plans to recruit trained scientists to answer technical questions and provide user support. It should also devise a mechanism for tracking user requests and the corresponding DAAC responses.

Data Set Preparation

The panel is concerned that, based on current plans, the ECS will not provide capabilities for reformatting options and geographical subsetting prior to the AM-1 and Landsat launches. These issues are likely to be of considerable importance to many users. As noted above, many users do not have experience with HDF-EOS format. In addition, the very large data volumes necessitate the capability for geographical subsetting because most users will not have the facilities or the bandwidth to download the full data sets. The latter applies particularly to global MODIS data sets. The panel therefore recommends that the DAAC develop plans for providing these formatting and subsetting capabilities, perhaps in consultation with other DAACs facing the same issues.

Recommendation 5. In the absence of an ECS capability, the DAAC should develop means for geographical subsetting and alternative formats in preparation for distributing Landsat 7, ASTER, and MODIS data.

Responding to New User Needs

It is important for the DAAC to continually reach out to serve new users. For example, before the launch of Landsat 1, an announced goal was to make data available on request within three days of their acquisition. This was intended to serve a number of applications for which rapid perishability of the data's value was of key importance. This goal was never realized. The DAAC now has the opportunity to serve users who require data in near real time. These users are not currently a major component of the Landsat user community.

Charging for Data

The panel observed that the DAAC provides all data free of charge and does not charge even for the cost of the media. In the EOS era, there is likely to be considerable demand for the data from a broad range of users in addition to scientists. This demand may create a backlog and make it difficult to provide the data to users in a timely fashion. The panel therefore concurs with a recent decision by the DAAC to charge for the costs of delivery and reproduction of popular data sets (e.g., GTOPO30). This decision would likely encourage users to order only the data they need.

TECHNOLOGY

Version 0

Because the EDC DAAC has no heritage data sets, it has bought little hardware and done little development. Rather, its current focus is on data distribution, which is carried out through Version 0, an information system developed jointly by all the DAACs. The Version 0 system—hardware and software—will be phased out after the ECS is delivered. In the meantime, the EDC DAAC uses Version 0 to distribute several large data sets, including GTOPO30 elevation data, global AVHRR 1-km data, and NASA Landsat Pathfinder data. The panel viewed a demonstration of Version 0 and was impressed by the capabilities of the user interface, but felt that components of the system could be improved through user feedback.

Hardware Availability

In the EOS environment, the ECS contractors will supply the people, development, and hardware needed to archive and distribute the data. The complement of ECS equipment supplied to the EDC DAAC is extensive, and includes SGI and Sun processors, 3.3 TB of disk storage to support the processors, and STK and EMASS tape archive units. The hardware delivered by the ECS contractor is clearly adequate to handle the large data sets that the DAAC will soon be managing. However, the hardware, which was delivered in spring 1997 and sits idle, could become obsolete, probably in a few years. The EDC DAAC has to develop a long-term hardware plan that includes incremental upgrades during the lifetime of the project.

Processing Software

The EDC DAAC is concerned that the ECS will not have the capacity to create products, particularly MODIS products, to meet user demand. At the time of the review, the DAAC had not seen the ECS software, and even the ECS liaisons were uncertain about the software because some elements were being installed remotely. Six months later, the DAAC had begun to work with the ECS software, but found that significant, required capabilities of the system had not been delivered and that the stability of the system for processing data was poor. NASA apparently shares these concerns, and it now seems likely that the ECS will not be used for processing data from the AM-1 platform.

Media Versus Web Distribution Strategy

All Version 0 data (GTOPO30, AVHRR, SIR-C, Landsat Pathfinder, and aircraft scanner data) can be accessed through one of several Web pages for the DAAC. However, the data are mostly stored off-line, and the DAAC seems reluctant to move them on-line. To disseminate data on-line, the DAAC would have to develop tools to subset the large data sets and to enable users to download the resulting smaller data sets over the network. In the absence of such tools, data are mostly delivered on media, rather than over the network. In the panel's view, growing numbers of users will want to access data over the Web, and the development of subsetting tools (see Recommendation 5) should go a long way toward meeting this need. The panel recognizes that developing a subsetting capability is a tall order for large data sets and that the DAAC may have to meet this challenge by offering specific, restricted subsetting schemes and by restricting the volume of downloaded data per user.

Connection to the World

The DAAC is connected to the ECS through the NASA Science Internet (NSI) router, which then connects to the ECS router by way of the Fiber Distributed Data Interface (FDDI). This connection has a bandwidth of 100 mbps. To connect to the Version 0 system, the NSI router links with the external router of the EROS Data Center. This connection is predominately Ethernet, which has a bandwidth of 10 mbps. External users access both the Version 0 and the ECS systems through a T1 link, which has a bandwidth of 1.5 mbps. Consequently, the T1 link is the bottleneck for remote access, and the DAAC plans to upgrade to increase bandwidth by the fall of 1998. Given that most data are delivered on media, current network bandwidth exceeds demand. However, low demand for on-line data sets may well be due to the lack of subsetting tools discussed above and, possibly, limits on bandwidth.

MANAGEMENT

General Philosophy

The DAAC manager, Lyndon Oleson, feels that the DAAC has little or no control over its fate—all decisions are made by ESDIS and the ECS contractor. Consequently, the DAAC has not yet formulated a clear vision for its future, but nonetheless seems optimistic that it will be able to fulfill its responsibilities. The DAAC offered no strategy for key elements of its operations, such as those involving scientists, characterizing the user community, subsetting data sets, and establishing performance measures. Without well-defined performance measures, the DAAC runs a risk of coming up short in the completion of its mission.

Recommendation 6. The DAAC should create a vision of its future role in EOS and other earth science missions and draw up an associated strategic plan, including concrete measures of performance to achieve this vision.

Personnel

The EROS Data Center complex, which includes the DAAC, is a matrixed operation, and most staff work part-time for the DAAC as needed. Currently, some DAAC staff are working on ECS maintenance and operations. The DAAC hopes to acquire 50 more ECS staff in preparation for the Landsat and AM-1 launches, although this may not be necessary if the data are not processed at the DAAC. If and when the ECS contractors arrive, DAAC staff who currently work on Version 0 and ECS maintenance and operations will be reemployed by the EROS Data Center.

ECS contractors, including the ECS science and engineering liaisons, are not integrated into the DAAC but are kept separate from DAAC operations. They will receive a limited amount of training on DAAC issues, but most are expected to already have skills such as system administration and software development. A discussion with the liaison and the head of maintenance and operations did not reveal any major concerns among the ECS contractors.

Budget

The DAAC's budget has grown from approximately $3 million in FY 1994 to $6.5 million in FY 1998, and is projected to reach $17.3 million by FY 2002, the latest year for which budget figures were available to the panel (Table 5.1). ECS hardware, software, and personnel account for about 30% of the current budget. The growth in the DAAC's budget is due partly to hardware acquisitions (e.g., FY 1997), and partly to an increase in operations and maintenance (e.g., FY 2001).

To illustrate its cost-effectiveness, the DAAC gave an example of how it had saved time and money developing Version 0 by leveraging off existing EROS Data Center capabilities. Producing the ASTER digital elevation model was also a good learning experience that will pay off for the DAAC in the AM-1 environment. The panel feels, however, that the DAAC would benefit by devising more quantitative measures of cost-effectiveness and tracking them over time. This is particularly important in an era of fiscal pressure on NASA budgets.

Contingency Plans

As noted above, NASA is likely to ask the DAACs or the science teams to process data from the AM-1 platform. Details on the MODIS, ASTER, and Landsat 7 science teams' contingency plans were not available to the panel, but the EDC DAAC's plan is to write the data to tape and then to inventory and shelve the tapes. Only the science teams would receive data in quantities suffi-

TABLE 5.1. Total EDC DAAC Costs (million dollars)[a]

	Fiscal Year								
	1994	1995	1996	1997	1998	1999	2000	2001	2002
EDC DAAC	3.3	3.6	4.0	3.8	4.4	4.3	4.3	10.9	11.8
ECS hardware	0	0.4	0.4	11.1	0.4	5.3	4.8	1.0	0.9
ECS software	0	0	0.2	0.5	0.2	0.2	0.1	0.1	0.1
ECS personnel	0	0	0.1	0.2	1.5	4.3	4.5	4.4	4.5
Total cost	3.3	4.0	4.7	15.6	6.5	14.1	13.7	16.4	17.3

[a]Budget numbers for FY 1994-1997 are actual values; numbers for FY 1998-2002 are projections, as of May 1998.
SOURCE: ESDIS.

cient for calibration and validation activities. In this case, Level 2 MODIS data would likely arrive at least a year late from the GSFC DAAC (see Chapter 1, "*Recent Developments*"). In the panel's view, failure to deliver data to users would constitute a failure of the fundamental mission of the DAAC, and it is incumbent on the DAAC to work with the science teams to provide at least the processing and distributing functionality scheduled to be delivered in the ECS.

Recommendation 7. The DAAC should develop a contingency plan for delivering Landsat 7 and ASTER data to users in the event of inadequate performance of the ECS.

EDC DAAC AND THE EARTH SCIENCE ENTERPRISE

Relation to EROS Data Center

The DAAC is hosted by the USGS EROS Data Center, which provides the facility (NASA paid for the building extension that houses the DAAC) and most of the DAAC staff. The EROS Data Center will also provide the long-term archive of data from the EDC DAAC. NASA funds 98% of the DAAC's operational budget, and the USGS does not try to constrain the operations of the DAAC in any way.

The home pages of both the EDC DAAC and the EROS Data Center are linked, and users can access data from previous Landsat missions through the DAAC. The transition from DAAC to EROS Data Center home pages is seamless, and DAAC users must take care or they will order expensive data from the EROS Data Center, thinking that they are ordering freely distributed DAAC data.

Relation to ESDIS

The DAAC perceives ESDIS as being concerned exclusively with development. According to the EDC DAAC, ESDIS defines success as the delivery of the ECS, not the successful operation of the DAACs. Consequently, ESDIS discourages the DAAC from bringing up possible problems with the ECS that would lead to further slips in the schedule. The DAAC is also concerned about shifting responsibilities between Goddard and NASA Headquarters, which has led to confusion at the DAAC and a lack of continuity at ESDIS. Finally, the DAAC blamed ESDIS for several issues of concern to the panel, such as having no scientists on staff, not developing formats and subsetting tools that are suitable to the land processes community, and not considering the needs of applied users.

The panel agrees that a more operational focus at ESDIS is desirable, particularly since the first EOS satellite (TRMM) has already been launched. The panel also understands the need for the DAAC to satisfy the ESDIS requirements, which are intended to enable the DAACs to operate together as a system and to keep costs down. However, the most effective data centers go beyond the stated requirements, and the panel encourages the DAAC to take the initiative in meeting the specialized needs of its user communities.

Relation to Other DAACs

The EDC DAAC has a special relationship with the GSFC and ORNL DAACs. The GSFC DAAC will provide Level 2 MODIS data, which will be used by the EDC DAAC to create Level 2+ MODIS land products. The ORNL DAAC holds laboratory and field data, which complement the EDC DAAC's remote sensing holdings. Interactions with the other DAACs, however, are weak. Although it is recognized that the DAACs need to coordinate their activities, the EDC DAAC does not view the activities of the other DAACs as being relevant to its operations. Consequently, the DAAC is missing an opportunity to learn about new tools and techniques from the other DAACs. In particular, it seems that a closer relationship with other DAACs dealing with data relevant to land processes (e.g., NSIDC and ASF) could be beneficial.

Relation to ECS Contractor

The DAAC is counting on the ECS for its operations, but it did not have input into the ECS development and does not know what capabilities the ECS will eventually provide. The long-term ECS development is dealt with as a contract, not a partnership. The DAAC feels frustrated because it has little or no control over the ECS but must nevertheless implement and operate the system.

The ECS liaison has worked to convince the ECS contractor that the DAACs should be involved in decisions, especially on priorities. There are now some review boards with DAAC membership, and the ECS contractor is beginning to shuffle its deliverables and schedule to accommodate users.

SUMMARY

The EDC DAAC is a world-class organization with a number of accomplishments in its six-year history. Particularly impressive to the panel was the DAAC's preparation and distribution of useful land processes data sets, such as AVHRR and GTOPO30. A valuable resource today, these data sets also demonstrate the DAAC's ability to handle the large data sets that will result from the upcoming EOS missions. The panel was also pleased by the DAAC's good relationship with its Science Advisory Panel, which bodes well for the DAAC's ability to meet the evolving needs of its scientific user community. On the other hand, the DAAC's future user community will likely be dominated by Landsat applications users. To serve their needs, the membership of the Scientific Advisory Panel will have to be adjusted so that it includes commercial and governmental as well as scientific users.

As the EDC DAAC moves into the EOS era, it faces a number of challenges. Its greatest challenges, completing preparations for managing data from Landsat 7 and AM-1 data streams, and developing contingency plans for the inevitable failures in the processing system, have become more tractable because of significant launch delays. Rather than becoming complaisant, the DAAC should use the extra time to complete its readiness exercises and clarify its vision for serving its potentially large, diverse user community. To serve its users well, the DAAC needs to know who its current user communities are, how they are likely to change as new data sets and products become available, and what specialized services (e.g., subsetting, near-real-time data) they need. Consequently, improving user services through better training, adding staff with scientific backgrounds, and incorporating feedback from users on a day-to-day basis should be a part of the DAAC's vision. Finally, the vision should include a strategy for upgrading hardware on a regular basis to prevent it from becoming obsolete.

Because the DAAC is part of a larger EOSDIS, it must serve the needs of both the land processes and the broader earth science communities. EOSDIS was designed to foster interdisciplinary research, and for this goal to succeed, many nonspecialized researchers will use the data and products of the EDC DAAC. In the past, the DAAC has worked with ESDIS and the ECS contractor to customize the ECS to meet the specialized needs of the land processes community. Now that the ECS is nearing completion, the DAAC should work with its partners—ESDIS, the ECS contractor, and the other DAACs—to create EOSDIS.

6

Alaska SAR Facility DAAC

Panel Membership

J.-BERNARD MINSTER, *Chair*, Scripps Institution of Oceanography, La Jolla, California
FERRIS WEBSTER, *Vice Chair*, University of Delaware, Lewes
KURT L. FEIGL, Groupe de Recherche de Geodesie Spatiale, Toulouse, France
ANTHONY C. HEARN, RAND Corporation, Santa Monica, California
MICHAEL S. LOUGHRIDGE, NOAA National Geophysical Data Center, Boulder, Colorado
JOHN M. MELACK, University of California, Santa Barbara
ROBERT S. PRITCHARD, IceCasting, Inc., San Rafael, California

---

ABSTRACT

The Alaska Synthetic Aperture Radar (SAR) Facility (ASF) DAAC is located at the University of Alaska in Fairbanks. Its mission is to process, distribute, and archive SAR data collected at present exclusively from foreign spacecraft—the European Remote Sensing Satellites 1 and 2 (ERS-1, 2); the Japanese Earth Remote-Sensing Satellite-1 (JERS-1); and the Canadian RADARSAT-1. As specified in Memoranda of Understanding between NASA and the foreign space agencies, only limited quantities of data are acquired by the Alaska SAR Facility and distributed to NASA-approved investigators. The data that are the most accessible are largely from the Alaska and McMurdo station masks, with the result that polar researchers are usually satisfied with the DAAC, whereas researchers in other disciplines often are not. In addition, the authority and budget for the development of software and hardware necessary for the ASF DAAC to succeed is not vested with the DAAC itself, but with the Jet Propulsion Laboratory (JPL) developers. However, responsibility for satisfying users rests with the ASF DAAC. A key recommendation is therefore that NASA grant authority for the operation and development of the facility to the DAAC. Finally, the demand for SAR data from U.S. researchers far exceeds the supply, and the panel urges NASA to formulate and help implement a long-term national policy for the acquisition, processing, and use of SAR data for civilian purposes (i.e., research and commercial operations).

INTRODUCTION

The Alaska Synthetic Aperture Radar Facility was established in a Memorandum of Agreement between the University of Alaska and NASA in 1986. Its mission is to establish, operate, and maintain a receiving, image processing, analysis, and archiving facility for SAR data, which are collected exclusively by foreign spacecraft. JPL designed and installed the data acquisition and management system, and SAR data have been acquired and distributed since 1991. The ASF DAAC was created in 1990, and it now handles the data processing, distribution, and archive for the Alaska SAR Facility. The DAAC's current holdings include data from ERS-1 and 2, JERS-1, and RADARSAT-1 missions.

SAR data are useful for applications ranging from sea-ice dynamics to volcanology to ecosystem change (Box 6.1). Consequently, the user community is small, but growing. Its growth, however, is hindered by data restrictions, which are specified by MOUs between NASA and foreign space agencies.

The ASF DAAC is unique within the EOSDIS system, not only because of its international character, but also because the processing information systems are being developed at JPL, rather than by the EOSDIS Core System (ECS) contractor. Because of delays in the ECS, JPL was tasked with developing an interim information system; the final system will be provided by the ECS contractor if sufficient funds are available. However, the JPL and ECS systems are not interoperable, so the ASF DAAC faces a difficult transition period several years from now if current ECS plans are adhered to. Finally, unlike most other DAACs, the ASF DAAC is currently managing large data streams. As such, it is the first DAAC to try to "drink from the fire hose," and its experience may well be a preview of what other DAACs will face.

The Panel to Review the ASF DAAC held its site visit on December 18-19, 1997. At that time, the management of the ASF DAAC was in transition. This transition has not yet been fully completed. However, many of the fundamental issues raised by the panel pertain more to NASA's long-term development, management, and use of the facility than to the DAAC's operation and will retain

BOX 6.1. Vital Statistics of the ASF DAAC

History. The ASF DAAC was created in 1990 to process, distribute, and archive SAR data, which have been acquired by the Alaska SAR facility since the 1991 launch of ERS-1.

Host Institution. Geophysical Institute, University of Alaska, Fairbanks.

Disciplines Served. Oceanography, volcanology, glaciology, wetlands ecology, forestry, and geology.

Mission. To provide the polar and earth system science communities with high-quality SAR data in a timely fashion, in support of research and operational investigations in the disciplines listed above.

Holdings. The DAAC currently holds about 110 TB of signal data and expects to acquire 115 to 145 GB of SAR data per day or 42 to 53 TB per year, and to process roughly half of that.

Users. There were about 400 unique users in FY 1997.

Staff. In FY 1998 the DAAC had 65 FTEs, 1 ECS contractor, 5 graduate students, 8 undergraduate student assistants, and 6 part-time positions.

Budget. Approximately $14.3 million in FY 1998 (including DAAC costs, JPL development costs, and ECS-provided hardware, software, and personnel), decreasing to $13.5 million in FY 2000.

their relevance beyond the transition period. The following report is based on the results of the site visit and e-mail discussions with DAAC staff in June through September 1998, a meeting with JPL developers in January 1998, and a meeting with Paul Ondrus (ESDIS) in April 1998.

HOLDINGS

The ASF DAAC is critical to NASA's Earth Science Enterprise and the U.S. Global Change Research Program because it is the primary source of SAR data for U.S. researchers. These data are essential for answering important scientific questions in a variety of disciplines, as well as for detecting and monitoring natural hazards. Further, SAR data become even more useful when combined with other remotely sensed and ground-based data. The DAAC's current holdings are listed in Box 6.2.

BOX 6.2. Current Data Holdings

- *European Remote-Sensing Satellites (ERS-1, 2)*—Complex SAR data (10-m resolution), full-resolution SAR images (30-m resolution), and low-resolution SAR images (240-m resolution) from 3,000-km-radius circles centered over Fairbanks, Alaska, and McMurdo station, Antarctica. Data are available beginning August 1991 from the Alaska mask and October 1995 from the McMurdo mask.
- *Japanese Earth Remote-Sensing Satellite (JERS-1)*—Complex SAR data (10-m resolution), full-resolution SAR images (30-m resolution), and low-resolution SAR images (240-m resolution) from a 2,600-km-radius circle centered over Fairbanks, Alaska, and other limited areas. Data are available from May 1992 to October 1998.
- *RADARSAT-1*—Complex SAR data (10-m resolution), standard beams; full-resolution SAR images (25-m resolution); low-resolution SAR images (150-m resolution); full-, medium-, and low-resolution (50-, 100-, and 400-m, respectively) ScanSAR wide (500-km-swath width); and full-, medium-, and low-resolution (50-, 100-, and 400-m, respectively) ScanSAR narrow (300-km-swath width) data from 3,000-km-radius circles centered over Fairbanks, Alaska, and McMurdo station, Antarctica. Significant coverage outside these masks is also available. Data are available beginning February 1996.
- *NOAA Advanced Very High-Resolution Radiometer (AVHRR)*—1.1-km-resolution images from selected areas within the Alaska mask from 1974 to 1991.
- *Alaska Landsat Quick Look*—80-m-resolution images of Alaska from 1972 to 1990.
- *Alaska High-Altitude Aerial Photography (AHAP)*—Images of Alaska from 1978 to 1986.

SOURCE: NASA (1998).

Formats

The ASF DAAC uses an international standard developed by the Committee on Earth Observation Satellites, rather than the EOSDIS standard (Hierarchical Data Format [HDF]-EOS), because SAR data are recorded as complex numbers, a data type not defined in the HDF-EOS standard.

Data Restrictions

The supply of SAR data to U.S. researchers is limited by two factors: the operations of data recorders on-board the satellites and MOUs between NASA and the flight agencies of Europe, Canada, and Japan. The DAAC has little control over either. All three MOUs restrict the DAAC from distributing data to unapproved users, and the DAAC is therefore hamstrung when trying to meet the demand of U.S. researchers.

Without an operational on-board recorder to capture the data, the satellite must transmit the data to a station on the ground. This requirement defines the station "mask," the area in which the antenna can track the satellite. For the ASF DAAC, this is a circular area of roughly 3,000-km radius around Fairbanks. Most of the DAAC's holdings image this area.

Neither ERS-1 nor ERS-2 spacecraft carry an on-board data recorder. The MOU between NASA and the European Space Agency (ESA) permits the DAAC to distribute ERS data within the Fairbanks mask to approved investigators. Outside this mask, approved U.S. users must order data from ESA at a cost equal to the processing cost. These costs are paid by the approved user. Unapproved users must pay the market price (about $1,600 per scene). In addition, the ESA has periodically made limited amounts of data available at no cost to U.S. researchers through announcements of opportunity. On the one hand, this process limits access to data by new investigators, and on the other hand, a number of principal investigators are now reaching their quotas.

Similarly, the MOU between NASA and the National Space Development Agency of Japan allows NASA-approved investigators to request JERS data within the Alaska station mask. Prior to the failure of the on-board tape recorder, which was turned off in August 1997, limited amounts of data from outside the Alaska mask were also available from the DAAC. The quantity of data that can be ordered depends on the allocation limit of the project, which is set in advance. In addition, some U.S. researchers have obtained JERS data free of charge by submitting proposals directly to the Japanese government. The JERS satellite ceased operations in October 1998.

For scientists seeking to obtain data outside the ASF mask without paying market prices, RADARSAT may be the most viable choice. RADARSAT is the only civilian radar satellite with a functioning data recorder on board. Approved users may order data outside the Alaska station mask from the appropriate ground station. These costs, which are paid from the DAAC budget (not by the approved user), may reach $1 million during this coming year. The total U.S. allocation for RADARSAT SAR time is 1,519 minutes per 24-day cycle, of which only 114 minutes of on-board recorder time is allotted to U.S. users. Here again, the MOU between NASA and the Canadian Space Agency restricts the ASF DAAC from distributing data to unapproved investigators. The ASF DAAC is also responsible for ensuring that the above limits are not exceeded. Thus, the U.S. allotment is a

TABLE 6.1. Data Processing Levels for the ASF

Data Level	Description
Raw signal data	Serial stream of ones and zeros as it comes down from the spacecraft
Scanned data	Raw signal data that have been scanned and entered into the ASF catalog
Level 0	Computer-compatible signal data at original resolution, time ordered, with all communications artifacts removed. Data are stored as discrete files
Level 1	Processed signal data at full or low resolution, time referenced, and annotated or calibrated with ancillary information, including radiometric and geometric calibration coefficients and georeferencing parameters

SOURCE: ASF DAAC.

precious resource, which is both oversubscribed (by a factor of four) and a source of contention.

Processing Strategy

SAR data are processed on demand to Level 1 (Table 6.1), even if the product has been processed before for someone else. (About 30% of JERS data ordered are requested more than once, and the percentage is higher for RADARSAT.) The DAAC adopted on-demand product generation because the RADARSAT upgrade, during which the system was reconfigured to support RADARSAT processing, did not include a Level 1 archive. Since then, the MOUs with the Japanese and European flight agencies have increased the data rates considerably (by a factor of 10, according to DAAC staff). As a result, only a small fraction of the DAAC's holdings are actually processed, which presents scaling problems for the DAAC as the demand grows. Furthermore, because of this practice, users cannot be guaranteed that the data ordered are exactly what they need and indeed cover a specific geographic area of interest. This processing policy therefore places a practical limitation on "data mining."

Recommendation 1. The DAAC should pursue development of a data mining capability through routine production of low-resolution, Level 1, georeferenced data products in near real time.

The DAAC has occasionally given high priority to large processing requests,

FIGURE 6.1. Mosaic consisting of more than 3,000 RADARSAT-1 images, which were Quick-Look processed by the Alaska SAR Facility and personnel from NASA's Jet Propulsion Laboratory and the Byrd Polar Research Center. This joint U.S.-Canadian project has yielded a new view of the ice-covered continent, which will enable quantitative analysis of the glaciology, geology and coastal processes of both East and West Antarctica. SOURCE: Richard R. Forster and Kenneth C. Jezek, Byrd Polar Research Center. Copyright Canadian Space Agency 1997.

such as the RADARSAT Antarctic Mapping Project. An initial step was to synthesize a Quick Look mosaic of the rim of Antarctica (Figure 6.1), which enabled verification of complete coverage and permitted preliminary research prior to complete processing. When the DAAC is engaged in such a large processing project, however, the processing requests of all other users are put on hold. Despite their scientific value, large processing projects are a guaranteed source of discontent.

Long-Term Archive

The ASF DAAC currently archives two copies of raw signal data, and will continue to archive them for 10 years past the end of ESA missions and for 15 years past the end of the Canadian mission. At present, there appear to be no plans to migrate the data to NOAA or to maintain them at the Alaska SAR Facility beyond the 10- or 15-year period following a mission. The issue of long-term custody of SAR data is particularly complex in this case because such a large fraction of the data is subject to restrictions imposed by foreign agencies, most of which are under pressure to recover their investment by charging a fee for the data.

The contents and structure of any long-term archive holding other data products are tightly coupled with the processing strategy. Should the DAAC abandon the on-demand processing strategy in response to increasing demand and choose instead to process all data to Level 0, then a Level 0 archive will have to be designed and implemented. In addition, a suitable plan will have to be devised to reprocess systematically all existing holdings. At the site visit, this possibility was mentioned, but no specific plan or budget for this task was presented to the panel.

USERS

Characterization of the User Community

Approved researchers are able to obtain SAR data from the ASF DAAC at below-commercial prices. Consequently, researchers in a variety of subdisciplines—sea ice, oceanography, glaciology, geology, geophysics, and land applications—are the primary users of the ASF DAAC. In addition, certain government agencies with operational needs (e.g., sea-ice monitoring within shipping lanes) constitute another group of important customers (see "*Special Processing*," below).

Users can be divided into in-mask and outside-of-mask users. As a consequence of the ASF's geographic location, in-mask users are mainly polar-ice researchers; researchers in most other disciplines must obtain data from outside the mask. The ASF DAAC also processes data from McMurdo station in Antarctica, which further enhances the polar focus of the DAAC. The polar focus, however, does not include Alaskan volcanoes. The panel was surprised that the Volcano Observatory, which is located in the same building as the DAAC, is not moving more aggressively toward using SAR data from the ASF DAAC to monitor volcanoes.in the Aleutians, especially with interferometry. The panel suggests that the DAAC encourage scientists at the Volcano Observatory to use DAAC services for this purpose.

User Working Group

The function of the ASF DAAC User Working Group (UWG) is to hold biannual meetings, prepare and maintain a five-year plan, and advise the DAAC on user needs. At the time of the review, the UWG had not met since the summer of 1996 and was being reconstituted. (A new UWG met in May 1998.) Consequently, the panel spoke only with the former chair of the UWG, Peter Mouginis-Mark. The previous UWG felt that the DAAC was not responsive to its suggestions, so members stopped attending the meetings. In particular, the UWG felt that the priorities of the DAAC should be on the production of data sets that the community wants, rather than on data acquisition, which is a primary ASF function but not a DAAC function. The panel concurs.

The panel strongly believes that an effective UWG must be established. Experience with such groups at other DAACs indicates that they can provide valuable advice and guidance from an interested segment of scientists. Ideally, the membership will span a wide range of disciplines and types of users. The effectiveness of the UWG depends on regular meetings and regular communication of DAAC activities in between meetings. Such communication should document DAAC data distribution, staff activities both locally and with larger EOS functions, and track progress on action items identified at meetings. Typically, the DAAC scientist and an elected member of the UWG co-chair the UWG and run the meetings. The meetings are best designed to elicit interactions rather than be passive information transfer sessions.

Relationship with the Scientific Community

As noted above, the DAAC serves users working on problems that fall either in its mask or outside its mask. This dichotomy has resulted in two levels of support and satisfaction among its user community: in-mask users are generally served well and are satisfied, whereas outside-of-mask users are usually not satisfied. Few outside-of-mask users receive the data they need on time, if at all. The absence of a working scheme for tracking data orders (see "*User Services*," below) only aggravates their impatience. Several dissatisfied users have even gone so far as to call the director of the host institution, the Geophysical Institute. The DAAC is quite aware of the problem, and the current reorganization is intended in part to solve it.

Because in-mask data are more readily available to study polar processes, scientists at the Alaska SAR Facility and the Geophysical Institute are understandably concentrating on polar studies. Many of these studies are regional in nature and do not necessarily require collaboration outside Alaska. The panel encourages the Alaska SAR Facility to take a more outward-looking approach to research and thereby rebuild a satisfied national and international constituency.

Special Processing

One of the most demanding in-mask users is the National Ice Center, which requires Level 1 data within six hours of acquisition to forecast sea ice in the

Arctic region. Its requirements for speed and volume are affecting other users, which may have contributed to the panel's perception of favoritism toward polar studies. The Antarctic Mapping Mission, which monopolized processing resources at the ASF DAAC for several weeks, also contributed to this impression. The Amazon Basin mapping effort, although not focused on polar problems, is another example of a large project that monopolized much of the DAAC resources and affected users.

User Services

The DAAC devotes nearly 10% of its budget and five people to User Services. It has conducted several user surveys, the most recent of which is excellent and includes information such as how scientists use ASF DAAC data in their research. It also includes a table of who received data products, how many products they received, and when they received them. These metrics could be tracked over time to help the DAAC measure changes in user satisfaction.

However, the survey and a brief visit to the user services group by the panel indicate that user services staff do not provide an adequate level of scientific and technical support to users. Improving horizontal communication between the divisions of the Alaska SAR Facility, particularly between scientists and the user services division, should help the user services staff to answer questions from the DAAC's customers.

The absence of a data request tracking and feedback system at the DAAC must be addressed soon. An on-line server should allow users to determine the status of their project at any time. Milestones such as scheduling, acquisition, scanning, and processing should be listed when completed. At present, users must first telephone the DAAC to request that data be collected and then telephone again, possibly many times, to obtain the processed data.

Recommendation 2. The DAAC should improve the scientific and technical capability of user services. In addition, a tracking and feedback system for monitoring the status of customer requests should be put in place, and statistics accumulated to provide metrics on the efficacy of this system.

Software Support for Interferometry

The panel praises the efforts of the science team at the Alaska SAR Facility to develop and distribute software for interferometric analysis of SAR data. This type of analysis has perhaps the greatest potential for fostering rapid progress in earth science studies because it is possible to measure crustal deformation and ice flow with a spatial sampling 10 to 100,000 times denser than previous surveying techniques with a comparable precision. Although several space agencies are actively developing such software, the Alaska SAR Facility is the only group willing to distribute source code for its working software.

The ASF software suite offered to users was supported with funds from the ASF DAAC. The majority of tools contained in the software can handle data from all four existing radar satellites, and the DAAC is committed to keep the software updated. This suite of programs was developed by the ASF science team. In addition, a SAR correlator and interferometry suite of programs was developed in consultation with Howard Zebker, formerly at JPL and now at Stanford University. The roots of the Alaskan software appear to lie in Zebker's version of the JPL Fortran code (circa 1996), which was not widely available to the scientific community outside JPL. The ASF package appears to be a major improvement because it is written in the more flexible C language and is cleaner and more portable. Indeed, the "processor" part of the software runs on the Cray T3D at the University of Alaska.

Several specialized software modules are restricted, however. As noted on the ASF Web page for interferometry software (http://www.images.alaska.edu/index.html), "the programs which make up the interferometry package are at a restricted ftp site. Restricted programs are denoted by (*R) following the name in the Description section. For information on becoming a registered user in order to access restricted programs, see the ASF Software Agreement . . . . Note: This concern may be relaxed by ASF upon completion of software classification by the U.S. State Department. NASA HQ has requested that we limit international distribution until this can be clarified." In view of the scientific value of the SAR processing software, the panel suggests that the ASF and NASA explore ways to improve access to scientific software.

Implications of Foreign Spacecraft Data

Some concerns about the services provided by the ASF DAAC to U.S. users stem fundamentally from the lack of a U.S. national policy or program for civilian SAR. The United States operates no satellite-borne SAR, and neither NASA nor the NSF seems willing to purchase data from the three foreign flight agencies that operate civilian radar satellites in volumes that would foster major scientific advances and at the prices quoted by these agencies. User demand far exceeds the amount of data available under the terms of existing MOUs. Unless NASA negotiates more favorable terms for U.S. researchers in future MOUs or purchases SAR data from foreign space agencies, the ASF DAAC will not be able to satisfy its user community any better in the future. For example, the panel heard numerous anecdotes that the Canadian Space Agency has placed severe restrictions on the amount of data U.S. researchers can obtain affordably—a situation that was not anticipated by many U.S. researchers. This unfortunate circumstance is compounded by the poor performance of the SAR processor for RADARSAT data (see "*Processing Software*," below).

Recommendation 3. The United States should formulate a long-term national policy for the acquisition, processing, and use of SAR data for civilian purposes. To implement this policy, NASA should create a focus (e.g., a program office) for the scientific use of SAR data.

TECHNOLOGY

Most of the ASF's production hardware and software is selected, developed, and installed by JPL. At the present time, it appears that JPL has the lead role in developing an agenda for facility development and that the ASF decides on the priorities within the resulting list. In talking to the ASF production staff, however, it became clear that the staff have the responsibility to make this facility work, but not the authority to ensure that it is working as effectively as possible. This issue is discussed in more detail below (see *"Relation to the JPL Developers*").

Hardware

The production facilities of the ASF consist of four components:

1. satellite data acquisition facilities,
2. data processing facilities,
3. data distribution facilities, and
4. data archiving facilities.

In evaluating the DAAC's responsibilities within the production facilities, the panel tried to ignore the first component because data acquisition is the role of the Alaska SAR Facility, not the ASF DAAC. Yet the line between data acquisition and processing and dissemination is somewhat artificial and can be drawn in different places according to what criteria are used (e.g., budget, personnel, hardware, data product level). Still, data acquisition appears to be the strongest part of the facility. Indeed, the panel was impressed with the ASF's success in tracking satellites at the 97-98% level.

With the exception of the satellite antennas, the four components are all housed in a single, cramped computer room. Although this is not necessarily a problem, it does mean there is little room for expansion or for the installation of alternative equipment during a transition phase. The equipment being used, which was selected or built by the JPL developers, is also quite heterogeneous. Although this too is not necessarily a problem, it does raise concerns about the maintenance costs associated with the disparate pieces of software across the various platforms. Finally, there exists a long-standing problem associated with the obsolescence of certain critical components, such as high-performance tape drives no longer supported by the original manufacturers, which can be maintained only by using custom-made replacement components.

One item of particular concern is the hardware processor, denoted the Alaska SAR Processor (ASP), which is a custom-built, special-purpose processor devoted to SAR data processing. As this equipment ages, the costs associated with the maintenance of both its software and its hardware will inevitably rise. In addition, the electronic components will not be replaceable before long.

Figure 6.2 illustrates the flow of data through the ASF processing system. Currently, raw signal data from the archive is converted to Level 1 products by the hardware and software processors. Neither processor can ingest data in Level 0 format. The hardware processor cannot easily be made year-2000 compliant, so it will be decommissioned on December 31, 1999. The software processor, on the other hand, will be modified to ingest both Level 0 and raw signal data.

At the time of this review, the archival data storage consisted of open tape racks that were accessed manually. The ECS contractor has since installed a StorageTek mass storage silo and the software to operate it, but the equipment, which was procured early because of a budget opportunity at ESDIS, will remain idle until the JPL developers procure a Level 0 processor and integrate it into the system, probably by the end of 1999. Consequently, the open tape racks will continue to serve as a data archive for at least the next several years. A prototype Level 0 processor has been procured, and the ASF DAAC, JPL developers, and ESDIS are currently determining how many will be necessary to attain a parallel processing capability that exceeds the raw signal data rate. The goal is to simultaneously carry out Level 0 processing of the raw signal data and complete migration of existing data from the open tape racks in the year 2000.

Processing Software

The uncertain health of the primary radar processor at the ASF and the current processing structure that employs one executable code for processing RADARSAT, ERS, and JERS data require serious attention by combined efforts in Alaska and at JPL. Specific improvements would be to provide 16-bit data and to process whole strips instead of single scenes for users requiring regional coverage.

Accessibility

Accessibility of the ASF DAAC's holdings is poor. The DAAC has no plans for distributing data on CD-ROM as the ESA now does, and the Web pages are largely informational instead of permitting convenient data access. Users should be able to browse images, download data sets and software, and order data and publications via the Web. Some of these features are available through the Version 0 Information Management System, but the interface is poor compared with

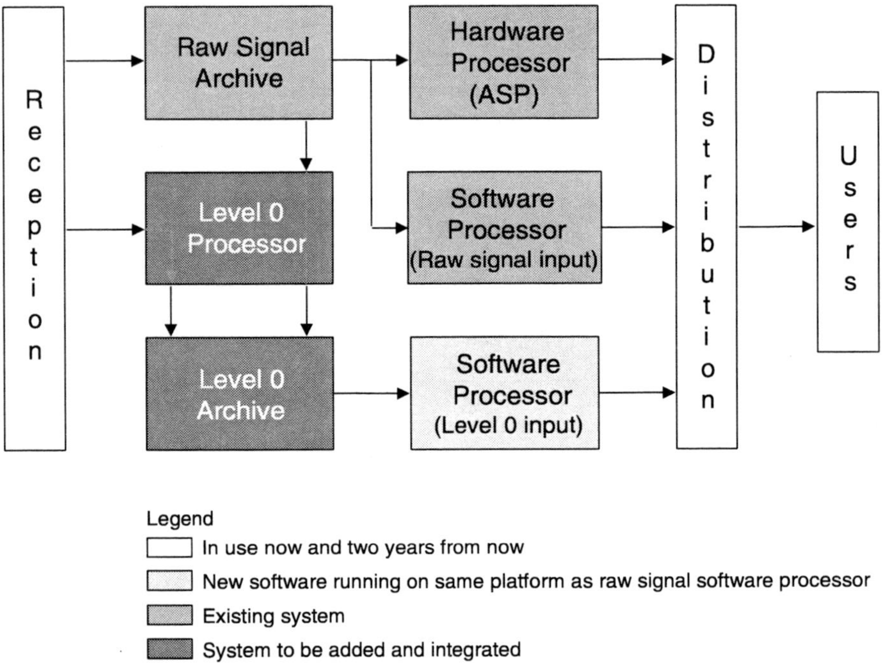

FIGURE 6.2. Schematic illustration of the flow of data through the ASF processing system. SOURCE: ASF DAAC.

modern Web technology. In addition, browse capabilities would enable timely access to SAR data, thereby allowing their use in near-real-time applications, such as sea-ice forecasting and volcano monitoring. The User Working Group has been unable to get the DAAC management to devote people and resources to the development of a fully functional Web interface, and the panel agrees that greater use of the Web would help build the DAAC's user community.

Recommendation 4. The ASF DAAC should develop an effective Web interface that enables authorized users to browse, order, and retrieve SAR data.

Connection to the World

One problem affecting the ASF's ability to communicate with its user community is the relatively limited bandwidth between the SAR facility and the rest of the world. All Internet connections are limited to T1 speeds (approximately 1.5 mbps), which limits the amount of data that can be conveniently exchanged. There are however, plans underway to install fiber-optic connections between

Alaska and the lower 48 states, possibly in 1998. This would provide for much higher bandwidth and thus reduce the current relative isolation of the SAR facility.

MANAGEMENT

This review took place during a time of transition, which is not yet complete. Syun-Ichi Akasofu, the director of the Geophysical Institute, has formed a faculty committee to evaluate proposals to reorganize the Alaska SAR Facility. The proposals have been evaluated, and a search committee has been formed to hire a permanent director, who will also serve as DAAC manager and chief scientist. The reorganization is scheduled to be completed in the fall of 1998. In the meantime, Craig Lingle, a faculty member at the University of Alaska, serves as acting director. The need to reorganize was apparently driven by dissatisfaction among many users, strain within the DAAC management team, and persistent tensions in the relationship between the ASF DAAC and the JPL developers.

Organization

The Alaska SAR Facility has two components: (1) a satellite-receiving ground station, which is responsible for data acquisition and antenna operation, and (2) the DAAC, which is responsible for data processing, distribution, and archiving. The DAAC component accounts for nearly 80% of the budget. The receiving ground station is funded by NASA via the Wallops Flight Facility. The DAAC is funded by NASA via the Goddard Space Flight Center. In practice, the Alaska SAR Facility and ASF DAAC operate as a seamless organization from the point of view of budget and personnel.

At the time of this review, the activities of the Alaska SAR Facility were carried out by nine divisions, including data management, data systems, engineering and maintenance, management, operations, planning, science, systems coordination and development, and user services. (At present, the organization has seven divisions.) In the panel's view, the large number of divisions presents a potential management problem by making it difficult to oversee the end-to-end operations of the DAAC. The panel also observed that the functional partitions between the divisions are firm and the boundaries between divisions are quite impervious, which inhibits sharing of technical resources and collegial or cooperative approaches to problem solving.

Personnel

At the time of the review, an acting DAAC manager had just been named, and it was difficult for the panel to evaluate personnel issues. Nevertheless, it was clear that the main tensions were between ASF DAAC staff and JPL developers, at both the management and the working levels (see "*Relation Between ASF*

TABLE 6.2. Total ASF DAAC Costs (million dollars)[a]

	Fiscal Year								
	1994	1995	1996	1997	1998	1999	2000	2001	2002
ASF DAAC	4.1	4.3	5.6	6.3	7.0	6.8	7.0	6.4	6.2
JPL component	6.2	9.2	10.3	8.1	5.3	5.7	6.3	4.8	3.5
ECS hardware	0	0	0	0.6	1.1	3.2	0.1	0.1	0.1
ECS software	0	0	0	0	0	0.4	0	0	0
ECS personnel	0	0	0	0	0	0	0	0	0
Total cost	10.3	13.5	15.9	15.0	13.4	16.1	13.4	11.3	9.8

[a]Budget numbers for FY 1994 - 1997 are actual values; numbers for FY 1998 - 2002 are projections, as of May 1998.
SOURCE: ASF DAAC budgets were supplied by Craig Lingle, ASF DAAC manager; JPL component budgets were supplied by David Nichols, ASF Development Project Manager; ECS hardware, software, and personnel budgets were supplied by ESDIS.

DAAC and JPL," below). Responsibility and authority are split between the ASF DAAC and the JPL developers, exacerbating the normal tensions between operations and development.

The DAAC's relationship with its ECS liaison, on the other hand, seems more positive. The ECS science liaison is treated as an ASF employee, rather than as an outside contractor. Consequently, she may become an effective bridge between the ASF DAAC and the ECS contractor if and when the ECS is delivered.

Budget

The ASF DAAC's total budget is $13.4 million in FY 1998 (Table 6.2). Development by the ECS contractor (8% of the budget in FY 1998) and the JPL developer (40% of the budget in FY 1998) accounts for approximately half of the budget. The ECS costs, which are largely associated with hardware and software acquisition and maintenance, will become insignificant by FY 2000. JPL development costs likewise decline from 68 to 36% of the DAAC's budget over the nine-year period shown. It should be noted that some of the JPL development efforts are directed toward the ground receiving station, which is not part of the DAAC. Therefore the costs of the JPL component shown below are maximum values. Overall, the DAAC's total budget will decline to $9.8 million by FY 2002, the latest year for which projections were provided to the panel.

The only measure of cost-effectiveness presented to the panel is the average cost of a Level 1 data product delivered to a user. It is calculated as the ASF budget averaged over some time period, divided by the number of images produced and distributed during the same time. This metric is misleading because the ratio of Level 1 products ordered to Level 0 products archived is quite low (less than 10%). Nevertheless, the cost per Level 1 product, measured this way, has dropped substantially due to recent major increases in processing volume, which led to a large increase in the number of images flowing through the system. The increases in processing volume were related to the debugging of RADARSAT processing code, system changes and upgrades, and management decisions on the RADARSAT processing workload and user priorities. As SAR imagery becomes more popular, these data will become increasingly important (and therefore valuable) over time.

In absolute terms, however, the budget of the DAAC is large compared with the small number of users (about 400). A high priority of the DAAC should therefore be to increase the size of its user community, thereby increasing its cost-effectiveness.

Strategic Plans

Quantitative metrics are a useful tool for measuring and monitoring the performance of a DAAC. Indeed, without them, it would be difficult for a DAAC to show that it has successfully fulfilled its goals. The ASF DAAC appeared to possess few metrics. At a minimum, metrics should be developed for determining how much data are acquired, how much are distributed, and how much are actually lost. (The panel heard informal stories about data losses that could not be documented or assessed.)

Another concern of the panel is that there appears to be no strategic plan for the evolution of the facility. The JPL developers sees themselves as providing "sustaining engineering and technology insertion." In other words, they do not see future planning as one of their responsibilities. One key issue is a replacement for the ASP, which, as noted earlier, will become increasingly expensive to maintain as time goes on. Some experiments by the ASF Science Division using the Arctic Region Supercomputing Center's Cray T3E are quite promising in this regard as an alternative. The JPL developer is also considering a "software architecture" for future developments. However, no one appears to be coordinating these parallel efforts.

ASF DAAC AND THE EARTH SCIENCE ENTERPRISE

Relation to Geophysical Institute

The ASF DAAC is considered to be an important part of the Geophysical Institute. The director, Akasofu, is quite aware of DAAC activities and instituted a reorganization and personnel changes to improve its operations. Akasofu is also seeking to improve the relationship between the ASF DAAC and the JPL developers.

Surprisingly, despite its importance to the Geophysical Institute, the DAAC has been allocated very little space for hardware. Space is tight, and new equipment such as the StorageTek archive will have to be located off-site. Unless the University of Alaska and the Geophysical Institute increase their commitment to the DAAC and allocate sufficient space for its operations, NASA may have no choice but to relocate the DAAC to more spacious quarters.

Several scientists from the Geophysical Institute attended the review, but the chief scientist and the acting ASF director are the only scientists on the DAAC staff. The DAAC's main interaction with scientists at the university is through the science division—about six scientists at the Geophysical Institute contribute regularly to the DAAC. On the other hand, the DAAC has acted as a catalyst for remote sensing research at the university, particularly the Geophysical Institute. The panel hopes that this productive interaction will continue in the future.

Relation to ESDIS

Paul Ondrus at ESDIS is responsible for the ASF DAAC and the JPL development contract. During an interview with the panel, he confirmed that JPL has the technical lead for the ASF DAAC and that JPL considers him to be the ASF DAAC manager. Before his arrival, several other people at ESDIS were responsible for the ASF DAAC. DAAC personnel have opined that the high turnover rate at ESDIS has led to a lack of continuity in long-term planning for the DAAC and may have adversely effected the DAAC's budget.

At a higher level, the DAAC feels somewhat ignored by NASA because it holds data from foreign spacecraft, rather than from NASA instruments or field experiments. There is, for example, no NASA project office that deals with SAR data.

Relation to Other DAACs

To be most useful, the data sets of the ASF DAAC should be combined with other types of data. Many of these complementary data will be provided by the other DAACs. For example, scientists using SAR data for interferometry will frequently request digital elevation models (DEMs) from the EDC DAAC, and scientists interested in improving the DEMs may want to use SAR data from the ASF DAAC. Similar examples abound in other disciplines (polar science, oceanography), and linkages with other DAACs (e.g., NSIDC DAAC, PO.DAAC) should be formalized to ensure that scientists can obtain the data they need. At present, the integration of other EOS data sets with SAR data held by the ASF DAAC is possible for sophisticated users, but the convenience of such a task is far from the standards of EOSDIS. As noted above, the DAAC is not on the delivery schedule for the ECS, and there are no plans to make the JPL and ECS systems compatible.

> **Recommendation 5. To make the ASF data sets useful to scientists in a variety of scientific disciplines, the ESDIS Project should integrate the ASF DAAC conceptually and technologically into the EOSDIS system.**

Relation to the ECS Contractor

Delays in the ECS have caused the ECS contractor to focus almost exclusively on those DAACs that will be receiving data from the AM-1 platform. The other DAACs, including the ASF DAAC, will not receive the ECS in the near future, if at all. When the ASF DAAC was on the ECS delivery schedule, DAAC personnel attended ECS planning meetings and participated in many phases of ECS planning and development. Funding for these activities was provided by ESDIS, over and above the DAAC budget. These activities have now ceased. In the panel's view, the lack of coordination between the ASF DAAC and the ECS will lead to increasing isolation of the SAR user community from the more integrated EOS community. It also affects improvements needed at the ASF DAAC, such as software links to the broader community.

Relation to the JPL Developers

Because the priorities of the ECS contractor have shifted to the AM-1 DAACs, ESDIS asked JPL to build an "interim solution" for processing data and to keep the system going until the ECS arrives. The JPL system, however, is incompatible with the ECS, and JPL developers argue that this is because they were neither required nor funded to make the two systems compatible or to make the ASF DAAC interoperable with the other DAACs. Moreover, ESDIS has apparently directed JPL away from exploring interoperability issues with the ECS contractors until after the ECS is delivered to the AM-1 DAACs.

ESDIS funds the ASF DAAC and the JPL developers under separate contracts. A five-year enterprise plan outlines the roles and responsibilities of each component. The JPL developers function as the instrument team for the DAAC because they provide Product Generation Executables (PGEs), a software element that accepts low-level data products and ancillary information and outputs a higher-level data product. They are also the equivalent of the ECS for the ASF DAAC.

The relationship between ASF and the JPL developers has apparently been strained for some time. From the DAAC's point of view, JPL's focus has been on optimizing processors, not on building an integrated system, as wished by the DAAC. Other DAAC requests for fixes are similarly ignored by JPL. From the

JPL developers' point of view, however, the DAAC asks for more than JPL is budgeted to provide—JPL considers the requests, but it can only act on the highest-priority items.

Panel visits to the JPL developers and Ondrus subsequently clarified the decision process: Ondrus sets the priorities for the JPL developers. The JPL developers have no plans to understand better the needs of the ASF DAAC by detailing staff to the DAAC for short periods of time, for example. The ASF DAAC hopes that the new five-year plan developed with ESDIS will alleviate the tension between the DAAC and JPL. The plan calls for the DAAC to establish operational needs and requirements, and for the JPL developers to meet these needs. In view of the poor performance of the SAR processor to date, NASA should evaluate the option of using alternate developers or fund the ASF DAAC to do system development in-house.

The panel views this structure as one that perpetuates a classical management problem arising when authority is not vested in the same persons or organizations that are burdened with the responsibility. In this instance, since the ultimate mission of the DAAC is to satisfy its users, this point of view is inevitably that the DAAC should be given the appropriate authority to discharge this responsibility.

Recommendation 6. The current contractual relationship between the ASF DAAC and the JPL developers is not responsive to the needs of the DAAC or its users. NASA should rework this relationship so the authority for the operation and development of the facility is vested in the ASF DAAC. Only then can the DAAC discharge its responsibilities to its users. This might require NASA to look elsewhere for a developer that meets the needs of the ASF DAAC.

SUMMARY

Synthetic aperture radar is a spectacular technique for studying the Earth. At present (except for a few airborne and shuttle-based experiments of limited duration), it is collected exclusively by foreign space agencies, which, by agreement with NASA, allow limited amounts of SAR data to be acquired and distributed to NASA-approved investigators at below-commercial prices. The lack of a national SAR policy, however, has led to problems in data acquisition. Nevertheless, as the primary source of affordable SAR data to U.S. researchers, the ASF DAAC provides an important resource. The DAAC's role in EOSDIS is almost equally important to researchers because SAR data are most useful when they are integrated with other types of scientific data. By serving as part of an EOSDIS system, rather than as an independent data center, the ASF DAAC has the potential to foster the type of multidisciplinary research for which EOSDIS was designed.

As illustrated in its excellent user survey, the ASF DAAC has had mixed success in meeting the needs of its scientific user community. Scientists who need data from the Alaska station mask tend to have good interactions with the DAAC and to be able to obtain the data they need. These scientists, many of whom work at the Geophysical Institute, concentrate largely on ice motion studies. Some of these studies require access to near-real-time data, and the ability of polar scientists to obtain them bodes well for scientists seeking to monitor volcanoes. Other studies, such as those supported by the Antarctic Mapping Mission, require a large amount of processing time. The impressive Quick Look product of the rim of Antarctica demonstrates the DAAC's ability to "drink from the fire hose," a challenge that the other DAACs will not face until the launch of the EOS satellites.

On the other hand, scientists who need data from outside the Alaska station mask are commonly frustrated, mostly because NASA has not concluded the necessary data acquisition agreements with foreign ground stations. Nor is it willing to fund U.S. researchers to purchase the data from foreign agencies at market prices. However, even if data can be acquired through the ASF, the DAAC has no tracking and feedback system for monitoring user requests. Consequently, users have no choice but to call the DAAC repeatedly to learn the status of their data acquisition and/or processing request. By implementing a tracking and feedback system, the DAAC will improve its relation with out-of-mask users, better serve the needs of the in-mask user community, and increase the size of its user community overall.

Another immediate challenge for the DAAC is to improve its relationship with ESDIS and the JPL developers. Currently, responsibility and authority are distributed among three organizations: (1) the ASF DAAC is responsible for operations and for satisfying user needs; (2) the JPL developers are responsible for creating the data acquisition and processing systems; and (3) ESDIS has authority for setting development priorities and overall budgets. This separation of authority from responsibility hinders the ASF DAAC from fulfilling its mission of serving the polar and earth science communities.

7

Physical Oceanography DAAC

Panel Membership

J.-BERNARD MINSTER, *Chair*, Scripps Institution of Oceanography, La Jolla, California
FERRIS WEBSTER, *Vice Chair*, University of Delaware, Lewes
DAVID DEWITT, University of Wisconsin, Madison
SYDNEY LEVITUS, NOAA National Oceanographic Data Center, Silver Spring, Maryland
LYNNE D. TALLEY, Scripps Institution of Oceanography, La Jolla, California

ABSTRACT

The Physical Oceanography DAAC (PO.DAAC) is hosted by the Jet Propulsion Laboratory (JPL). It manages data from a wide variety of ocean experiments and missions, including several done in collaboration with other countries, and it primarily serves the physical oceanography community. Although it has extensive experience with satellite data, the DAAC handles few data streams and will not be receiving data from the AM-1 platform. Consequently, the DAAC will have sufficient time to link its system with the EOSDIS Core System (ECS) (if and when it becomes available) and to scale up for future EOS missions.

The site visit showed that the DAAC is functioning well today, and it has the necessary strategic plans for operating successfully in the future. A primary

BOX 7.1. Vital Statistics of the PO.DAAC

History. The PO.DAAC was created from the NASA Ocean Data System in 1991. Its holdings go back to 1978.

Host Institution. NASA-Caltech Jet Propulsion Laboratory in Pasadena, California.

Disciplines Served. Oceanography and geophysics.

Mission. To make available to a wide user community data and information on ocean physics, and air-sea interactions, in easily usable form.

Holdings. The DAAC holds approximately 15 TB of heritage data sets and receives 4-5 TB of data per year. None of its future data sets will come from the AM-1 platform.

Users. There were 15,527 unique users, including 2,000 regular users, in FY 1997.

Staff. In FY 1998 the DAAC had 28 staff and 1 ECS contractor.

Budget. Approximately $4.4 million in FY 1998 (including DAAC costs and ECS-provided hardware, software, and personnel), increasing to $6.4 million in FY 2000.

reason for its success is its location within the physical oceanography research group at JPL, a mutually beneficial arrangement that helps the DAAC understand how its data are used and the needs of researchers. This close working relationship, however, is jeopardized by recent trends to outsource DAAC functions, and the panel's main recommendation is that JPL should keep the DAAC intact and collocated with the oceanographers.

INTRODUCTION

The Jet Propulsion Laboratory has been managing remote sensing data from the oceans since the Sea Satellite (SeaSat) Program in the early 1980s. These collective data activities formed the NASA Ocean Data System, which became the basis for the Physical Oceanography DAAC. The DAAC has existed since 1991 and is responsible for processing, archiving, and disseminating all of NASA's data related to physical oceanography (Box 7.1). The DAAC deals with data from many spacecraft, including several from foreign countries. Each instrument yields one discrete data set, except for the Ocean Topography Experiment (TOPEX/Poseidon) altimeters, which yield a data stream. PO.DAAC data volumes are much lower than those of the other DAACs, so the data management problem is relatively tractable. As with the ASF DAAC, the PO.DAAC has been dealing with spacecraft data for quite some time; none of the other DAACs has had as extensive experience with an active satellite program.

The DAAC manages about 15 TB of data from a variety of ocean remote sensing missions (see Box 7.2). Future missions, which will add 4 to 5 TB of data to the DAAC each year, include SeaWinds 0/QuickSCAT, SeaWinds, Advanced Microwave Scanning Radiometer (AMSR), and Jason-1. These missions, which are scheduled to be launched in 1999 or 2000, are being flown in collaboration with other countries—SeaWinds and AMSR with Japan, and Jason-1 with France. None of these missions are related to the EOS AM-1 platform, and the DAAC's greatest challenge will be to ensure that the data become fully accessible through EOSDIS.

The Panel to Review the PO.DAAC held its site visit on January 8-9, 1998. The following report is based on the results of the site visit and e-mail discussions between the panel and the DAAC manager in June and July 1998.

HOLDINGS

The scientific value of the data sets for which the PO.DAAC has primary responsibility (see Box 7.2) is inestimable. Major discoveries are being made each year with the TOPEX/Poseidon data, which provide the first nearly global long-term sea surface height information. The wind data from NSCAT (see Figure 7.1) are better than wind data from any other data set or model analysis, and the QuickSCAT results are anticipated to be similarly spectacular. Finally, the Special Sensor Microwave Imager (SSM/I) Pathfinder information enables components of the fluxes between the ocean and atmosphere to be calculated, and the sea surface temperature information is important to a wide variety of ocean and climate studies.

Formats

Hierarchical Data Format (HDF) is the primary format used by the PO.DAAC, but the DAAC also maintains other formats, such as ASCII, for personal computer and Macintosh users who are unable to access the UNIX libraries needed for HDF. Some heritage data sets, such as SeaSat, are not being put in HDF for data distribution and exchange.

HDF is a subset of the standard EOSDIS format, HDF-EOS. Currently, none of the DAAC's data sets are in HDF-EOS format, and the DAAC will have to transition to the use of HDF-EOS over time if its data sets are to be used successfully in conjunction with other EOSDIS data sets.

DAAC staff were concerned about future support for HDF-EOS. A number of potential problems loom. First, the quality of the version of HDF used by the DAACs is not as good as the versions produced by the National Center for

BOX 7.2. Data Holdings as of January 1998

• *TOPEX/Poseidon Merged Geophysical Data Record (MGDR)-B*—Global data from the U.S. and French altimeters for August 1992 to present.
• *NASA Scatterometer (NSCAT)*—Global or ocean coverage for August 1996 to present.
• *NOAA Advanced Very High-Resolution Radiometer (AVHRR) Sea Surface Temperature*—Global daily, monthly, and yearly data for 1987 to present.
• *Nimbus-7 Scanning Multichannel Microwave Radiometer (SMMR)*—Ocean data for July to October 1978.
• *Defense Meteorological Satellite Program (DMSP) Special Sensor Microwave/Imager (SSM/I) radiometers*—Global or ocean coverage for July 1987 to December 1991.
• *Sea Satellite (SeaSat) scatterometer and altimeter*—Global or ocean coverage for July to October 1978.
• Assorted CD-ROMs, including the *GOSTA Plus Atlas, Power Ocean Atlas for the Macintosh,* and *ATLAST for the PC.*

SOURCE: NASA (1998).

Supercomputer Applications (NCSA). Second, the ECS contractor is evidently slow to fix bugs and provide support. Third, this version is not compatible with HDF 5.0, the version currently being implemented at NCSA. As a result, the DAAC will have to depend on the ECS contractor for the long-term development and maintenance of HDF-EOS. In the panel's view, NASA should consider contracting with NCSA to assume long-term maintenance and development of HDF-EOS. The long-term goal would be for NCSA to merge the capabilities of HDF-EOS into a future version of HDF 5.0.

Documentation and Metadata

Examination of the PO.DAAC's Web site indicates that on-line data sets are well documented. The DAAC archives ancillary data along with the data sets, and provides read software and detailed data set guides and user manuals for each data set. In addition, a reference list, sometimes quite extensive, is supplied with each data set. Some of the metadata provided by the DAAC (often from the flight projects) is more detailed than is required by EOSDIS guidelines. On the other

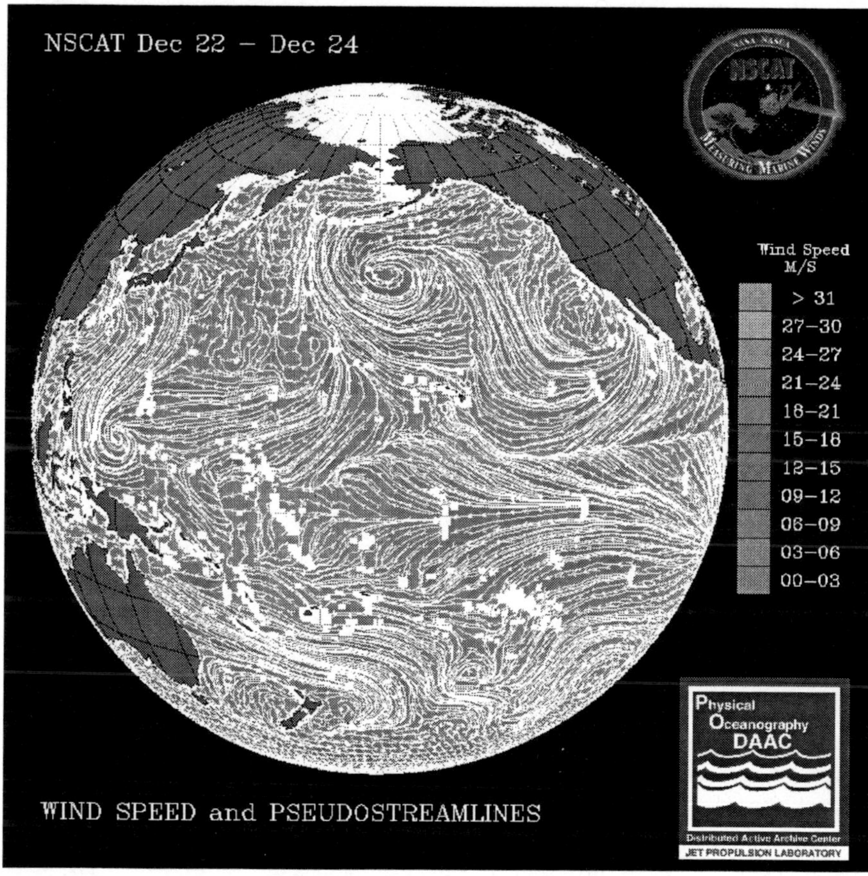

FIGURE 7.1. NSCAT-derived wind field for the period December 22 to December 24, 1996, showing both wind speed and pseudo-stream lines for the wind field. These data were taken by the NASA NSCAT instrument on the Japanese Advanced Earth Observing Satellite 1, and the image was produced by the PO.DAAC at JPL. SOURCE: PO.DAAC.

hand, documentation on the software is less complete, a problem acknowledged by the DAAC.

Data Restrictions

The PO.DAAC holds European Remote Sensing Satellite (ERS) data, whose use is restricted to "approved" NASA scientists under the terms of the MOU between NASA and the European Space Agency. The remaining holdings of the DAAC are available to users in any country.

Processing Plans

The PO.DAAC does not follow the EOSDIS model of processing data products using Product Generation Executables (PGEs) provided by the instrument teams. Most standard data products are produced by the flight projects, then transferred to the DAAC for distribution and archive. In this sense, the PO.DAAC operates more like a data center than a DAAC (see Chapter 2, "*DAAC Versus Data Center*"). Data products produced by the DAAC, such as the TOPEX/Poseidon Merged Geophysical Data Record, are not produced using instrument team PGEs. Rather, the DAAC develops the software and algorithms in-house, often with the oversight of members of the science team.

Reprocessing Strategy

The DAAC plans to reprocess some data sets annually because it is more cost-effective to rework data sets regularly than to ignore them for a long time. Reprocessing NSCAT and TOPEX/Poseidon data sets is among the DAAC's top three priorities for FY 1998. The panel commends the DAAC for its wise approach to reprocessing.

Subsetting Strategy

The DAAC has developed a tool that allows HDF files to be spatially subsetted. Different tools have been developed for different data types (e.g., point, grid, swath), and the Web interface allows a user to specify the desired spatial subset. This makes it possible to reduce the amount of unwanted data returned to a user. PO.DAAC staff were quite proud of this subsetting tool, which is not scheduled to be provided by the ECS contractor until 1999. The panel applauds the DAAC's initiative to subset HDF.

Long-Term Archive

Until NASA and NOAA develop a meaningful agreement on long-term archive of PO.DAAC data, the DAAC plans to maintain its data sets, provided that the budget and program continue on their present course.

USERS

Characterization of the User Community

The PO.DAAC's user community includes EOS science teams, ESDIS and other NASA program managers, data centers, physical oceanographers, K-12 educators, the private sector (e.g., fishermen), and the general public. The DAAC places its highest priority on serving the needs of the science teams (the main repeat users of the DAAC), the flight projects (the main data providers), and ESDIS.

The DAAC keeps statistics on its users, such as the number of U.S. and non-U.S. users of a particular data set, but the statistics are not compiled in such a way as to enable full characterization of the user community. Such characterization would help the DAAC better understand and serve the diverse needs of its users. It would also facilitate development of a more accurate EOSDIS user model, thereby improving user satisfaction with the system as a whole. Developing and tracking metrics on who its users are and how they use the data should help the DAAC achieve both objectives.

Recommendation 1. The PO.DAAC should develop a better profile of its user population, and gather statistics on DAAC accesses and data usage. These should be used to construct quantitative performance metrics to help ensure continued user satisfaction.

User Working Group

The PO.DAAC has a close relationship with its User Working Group (UWG). The UWG meets twice a year, once at JPL to provide input on the DAAC's annual work plan and once off-site to discuss broader issues such as priorities on data product development and data acquisition. The UWG is pleased with the DAAC's responsiveness to its recommendations, and current and former members with whom the panel spoke are strong supporters of the DAAC. At the site visit, UWG members pointed out several DAAC initiatives that they believe have made the DAAC successful.

Relationship with the Scientific Community

As noted above, the DAAC places a high priority on researchers who use and/or produce its primary data sets. Consequently, the DAAC's relationship with the scientific community is strong. Particularly impressive to the panel was the high level of interaction between the DAAC and the physical oceanography group at JPL, with which it is collocated. The DAAC scientist, Victor Zlotnicki, is a respected oceanographer, and this helps foster community trust in the DAAC.

Based on the large number of users (15,527), discussions with the User Working Group, and the CGED survey (Appendix D), it appears that the DAAC's relationship with oceanographers outside JPL is similarly strong. Indeed, it is not unusual for scientists to ask the DAAC to archive their data sets or develop software or tools. The panel notes that a close working relationship with researchers helps provide a scientific context for the DAAC's work, thereby helping it to fulfill its mission of facilitating research in ocean physics and air-sea interactions.

Scientists also provide an important and proven source of feedback and new ideas. In the panel's view, the DAAC's strong, ongoing interaction with scientists is a major factor in its success.

User Services

The DAAC believes that its real contributions are in the quality of the data produced and the scientific understanding that results from the use of these data. Consequently, user services are a high priority of the DAAC. It appeared to the panel that the DAAC is responsive to user requests for advice and suggestions, and a subsequent CGED survey (see Appendix D) confirmed this impression.

In addition, the DAAC has restructured its Web site and CD-ROMs to better serve its user communities. The Web site is easy to navigate and provides branches at the top that clearly direct various user groups appropriately (i.e., general public, educators, academic scientists). For example, the Web page feature on El Niño 1997-1998, which is kept up to date by the DAAC, is a stellar example of the use of Web technology to educate and inform the public at large. Similarly, the DAAC's educational CD-ROM products not only have provided a public service, but have also helped the DAAC expand its user base.

Foreign Access

The PO.DAAC is located within a secure facility, but the necessary clearances can be obtained when users need to visit the center for several days or weeks. For example, the DAAC recently hosted a Chinese graduate student who needed to visit for a few weeks to produce a data set. In addition, the Web provides access to all DAAC holdings, thereby reducing the need for physical visits to the center.

TECHNOLOGY

Hardware Availability

The DAAC's system is currently based on a combination of SGI and Sun processors. The tape library is a 4-TB Metrum RSS-48 unit employing VHS tape technology. At the time of the site visit, the Metrum tape library was being replaced with a StorageTek 9710 unit based on digital linear tape (DLT) technology. Each cartridge has a capacity of 35 GB (uncompressed). The unit purchased is configured with six readers and has a total storage capacity of 20 TB (uncompressed), which the DAAC estimates will be sufficient for both its existing data sets and the data sets produced by future missions for which it is responsible. Unitree is used as the DAAC's hierarchical storage system. Although other DAACs (e.g., GSFC) have evidently not been happy with Unitree, the PO.DAAC has found that with the proper amount of disk cache, it works quite well for the DAAC's data sets. Finally, the ECS contractor has purchased an EMASS unit for the DAAC, but the DAAC has no need for it or for the associated AMASS software, especially since the technology on which the EMASS unit is based is now obsolete.

The DAAC's investment in hardware is modest, and it seems to do a good job of bringing in new hardware in a timely fashion. Except for tertiary storage, the hardware is functionally partitioned by missions. That is, a dedicated computer (typically a small SGI multiprocessor) handles data product generation and processing requests for each data set maintained by the DAAC.

At the site visit, DAAC staff displayed an impressive awareness of the issues related to long-term archiving. Evidently, the task of restoring the SeaSat archive made clear how difficult a job maintaining a long-term archive is going to be. For a possible long-term archive media, the staff might explore the use of digital video disk (DVD) technology. As a "consumer" technology, it is likely that DVDs will exist for at least as long as CDs have (about 20 years). Although the archive media is only part of the problem (and, possibly, a small part), a stable archive media (i.e., a long period of commercial viability and a long shelf life so that the data remain readable over a long period of time) is a necessary requirement.

Processing Software

The DAAC software reflects careful and thoughtful engineering. Although some Level 0 and Level 1 data sets are maintained in their native format (or even in ASCII), the DAAC makes extensive use of HDF for storing data products (see Table 1.1 for a description of processing levels). The relational database system Ingres is used to keep track of data sets and data products. The archiving and product generation process is mostly a hands-off operation. Processing is controlled by a combination of PERL programs and shell scripts. The panel was impressed by the simplicity and effectiveness of the processing operation.

Because the Ingres product line is essentially in a maintenance mode, the DAAC will eventually have to switch to another database system. To minimize the impact of this eventual change when it occurs, DAAC staff have tried to avoid using Ingres-specific features to the maximum extent possible. When a switch occurs, the DAAC should consider the use of an object-relational database system with support for geospatial and temporal data (e.g., Informix Universal Server). This would significantly simplify the task of supporting requests for data sets covering a particular geographical region.

Media Versus Web Distribution Strategy

Although the PO.DAAC delivers some of its data sets and products via the Web, most large data sets are distributed using either CD-ROMs or tapes (8 mm or 4 mm). DLT tapes are much denser (up to 35 GB uncompressed data), but their relatively high cost is an impediment to widespread use. The DAAC is actively tracking DVD developments in the event that DVDs eventually become the standard distribution media.

In addition to DVDs, the DAAC is exploring the use of multicast and direct-broadcast satellite technology for distributing its data sets and products. Even though the use of direct-broadcast satellite technology to distribute large data sets is in its infancy, the Department of Defense (DOD) is making a major effort to use it to disseminate large geospatial, image, and video data sets. The DAAC should continue its exploration of the use of this technology since it may provide the most cost-effective distribution mechanism in the future. The panel would encourage the DAAC to consider a trial involving a commercial provider and a dozen key DAAC users.

Connection to the World

At present, bandwidth appears to be sufficient for the acquisition and dissemination of the DAAC's data sets. Beginning with the QuickSCAT mission, however, both internal and external network capabilities will have to be upgraded to handle the larger volumes of data. The DAAC is using fiber-channel and gigabit technologies to upgrade its local area network for the QuickSCAT data. In addition, it is working with Goddard Space Flight Center to ensure that its external networks have sufficient capacity to support the flow of data from the downlink at the Wallops Flight Facility to the QuickSCAT Ground System. Additional bandwidth will be needed in the future to support the SeaWinds and AMSR missions, and the DAAC has already requested a network review to ensure that it has the needed capabilities before launch.

MANAGEMENT

General Philosophy

The PO.DAAC's philosophy is to meet user needs by developing a data management system incrementally and from the bottom up. The panel agrees with this responsive and flexible approach to data management. The DAAC manager, Donald Collins, listens to people (users and DAAC staff), gets along well with the DAAC scientist, and encourages his staff to take chances. Morale is good at the DAAC, although the recent trend toward outsourcing may be beginning to have a negative impact (see below).

On the other hand, the panel found the DAAC to be somewhat introspective, with a strong focus on the physical oceanography community and only a passing interest in the needs of users from other disciplines. Although the panel agrees that the primary focus of the DAAC should remain on physical oceanographers, a broader multidisciplinary outlook could only enhance the position of the DAAC within the earth science community.

Personnel

Staffing is one of the largest problems facing the PO.DAAC. The DAAC finds it difficult to retain staff or fill vacancies, in part because salaries offered to engineers and computer scientists by JPL are not competitive with salaries for similar positions in the greater Los Angeles area. In addition, ESDIS recently cut the DAAC budget and the DAAC chose to absorb the cut by leaving vacant positions unfilled. The number of staff was further reduced in March 1997, when JPL management decided to outsource some DAAC functions as part of an effort to reduce the total number of JPL employees (see "*Relationship with JPL*," below). Thus far, all programming and database staff, half of the user services staff, and most of the operations staff have been outsourced and moved off campus.

Another personnel issue has to do with the classic tension between operations and development. Although tensions were marked when the system was first being developed, the DAAC has partly solved the problem by separating the hardware for operations and development.

Finally, the DAAC has historically had a good relationship with its ECS liaisons and considers them DAAC staff. In fact, the two previous ECS science liaisons were hired by the DAAC.

Budget

The PO.DAAC's budget grew from $3.3 million in FY 1994 to $4.4 million in FY 1998 and is projected to reach $6.5 million by FY 1999 (Table 7.1). The growth in the budget reflects increases in operations and maintenance, and the acquisition of ECS hardware (e.g., FY 1997). Except for hardware acquisition years, the ECS-provided hardware, software, and personnel generally amount to less than 10% of the DAAC's budget. The overall development effort is anticipated to be less than 30% of the total budget for FY 1998 to FY 2002, the only years for which detailed DAAC-specific budgets were available.

The DAAC offered the panel several examples of its cost-effectiveness. First, the number of files, volume of data, users, and requests for data have all increased over the past three years, but costs have decreased, indicating that the DAAC has become more efficient. Second, building a system incrementally is cost-effective. Third, as the SeaSat data restoration shows, reworking data constantly is ultimately more cost-effective than ignoring them for a long time. Finally, the DAAC plans to adopt only part of the ECS, rather than the entire system (see "*Strategic Plans*," below), which may lead to large cost savings.

TABLE 7.1. Total PO.DAAC Costs (million dollars)[a]

	Fiscal Year								
	1994	1995	1996	1997	1998	1999	2000	2001	2002
PO.DAAC	3.3	3.2	3.9	4.4	4.3	5.1	5.8	6.1	5.5
ECS hardware	0	0	0.6	3.5	0.1	1.2	0.4	0.2	0.1
ECS software	0	0	0	0.4	0	0.1	0	0	0.1
ECS personnel	0	0	0	0	0	0.1	0.2	0.2	0.2
Total cost	3.3	3.2	4.5	8.3	4.4	6.5	6.4	6.5	5.9

[a]Budget numbers for FY 1994 - 1997 are actual values; numbers for FY 1998 - 2002 are projections, as of May 1998.
SOURCE: ESDIS.

Strategic Plans

The ECS is one of the largest problems facing the DAAC, partly because of uncertainties in the modularity of the system and partly because the PO.DAAC is not on the near-term delivery schedule. The latter makes it difficult for the DAAC to make plans and to obtain accurate application program interfaces (APIs).

The DAAC has demonstrated that it can serve its users without the ECS. Linking its systems to the ECS will create work, but has several advantages. For example, the DAAC would like to use the ECS interoperability server to advertise its data sets. (The DAAC is doing a feasibility study to see whether it can adopt only this part of the system, and staff are confident they will succeed.) The DAAC also plans to use parts of the ECS for the Jason-1 data sets, particularly the ECS archive system and the means to distribute the data. The panel agrees that the DAAC's strategy of adopting only the parts of the ECS that it needs is sensible, given the context. However, by employing such a strategy, the DAAC risks losing coherence with the EOSDIS system.

Recommendation 2. The PO.DAAC should continue its strategy of adopting only those elements of the ECS that it needs. ESDIS should not oblige it to adopt all elements of the ECS.

Recommendation 3. ESDIS should ensure that the PO.DAAC obtains accurate application program interfaces and enable it to build its own modules to interface its system with the ECS.

PO.DAAC AND THE EARTH SCIENCE ENTERPRISE

Relation to the Jet Propulsion Laboratory

Organizationally, the PO.DAAC is located within the Information Science Division at JPL, but it is physically embedded in the Physical Oceanography Research Element. The integration and feedback that have resulted have been mutually beneficial, and the panel felt that this has resulted in a center that serves its scientific community well. Unfortunately, this close working relationship may not last, in view of JPL management's decision to reduce staff by outsourcing some DAAC staff and moving them off campus. As a result of this decision, only a fraction of DAAC personnel would remain at the JPL site and be able to interact daily with researchers in oceanography. Although the use of contractors to perform various services is common at JPL (and, indeed, at other DAACs), the DAAC's perception is that this solution does not result in any savings and leads to morale problems and a less efficient operation. A subsequent e-mail discussion with Diane Evans, program scientist in JPL's Earth Science Program Office, suggests that morale has improved now that a contractor has been selected (Raytheon). Moreover, Charles Elachi, director of the Earth and Space Sciences Office of JPL, told the panel that he is committed to supporting data management activities and ensuring that the PO.DAAC meets the needs of the community. However, the panel's main concern—that outsourced DAAC staff will be physically separated from the JPL oceanographers—remains an issue.

Recommendation 4. ESDIS and JPL management should keep the PO.DAAC intact. The DAAC staff and JPL oceanographers work well together. Physically dividing DAAC staff from the scientists by outsourcing will reduce the team's effectiveness.

Relation to ESDIS

According to Collins, the DAAC managers have been trying to strengthen their relationship with ESDIS. Although ESDIS knows it has communications problems with the DAACs, the DAACs believe that ESDIS' main communication problems are with the ECS contractor.

At the time of the review, the PO.DAAC was worried that the departed DAAC system manager, Gregory Hunolt, who had an operational focus, would be replaced with someone with a development focus, who would therefore not be philosophically attuned to the DAACs. The earlier departure of Dixon Butler, former operations director of the Data and Information Systems Division of Mission to Planet Earth, who displayed strong leadership and a vision for EOSDIS, was also a cause of concern to the DAAC.

Relation to Other DAACs

The DAACs coordinate certain activities, such as priorities, procedures, and protocols, and they seem to communicate well at working levels (in particular the User Services Working Group and, to a lesser extent, the Operations Working Group and Systems Engineering Working Group). Greater cooperation with other DAACs, however, is not a high priority for the PO.DAAC, mainly because there are no instrument interdependencies. The panel, on the other hand, notes that communication and interoperability with other DAACs are necessary for helping users locate and combine disparate data sets, thereby fostering multidisciplinary research. This is a major objective of EOSDIS.

Recommendation 5. The PO.DAAC should strengthen communications with other DAACs, especially with respect to interoperability issues.

Relation to the ECS Contractor

The DAAC has participated in ECS design reviews, but feels that the ECS contractor has not been responsive to its specialized needs. The DAAC's perception is that the ECS contractor is building to the specifications of the contract, which can be modified only by ESDIS. The ECS contractor has now stopped soliciting input for development and is focusing on delivery of the system to the AM-1 DAACs. Thus, the opportunity for input has passed.

SUMMARY

The PO.DAAC is a well-run, well-functioning data center that has provided value-added services (e.g., production and distribution of data sets, development of tools and algorithms) to the oceanographic community since the early 1990s. Its success can be attributed to several factors, including (1) a vision of doing whatever is necessary to satisfy the needs of its users; (2) a sensible strategy for keeping its data sets active through regular reprocessing; (3) a flexible, incremental approach to system development; and (4) a commitment to long-term archiving. The first may well be the most important, and the DAAC could serve its users better if it developed and tracked statistics on what and how data are being used.

The DAAC has an excellent relationship with its highest-priority users, physical oceanographers. This relationship is strengthened by the influence of the User Working Group in the DAAC's activities and by the collocation of DAAC staff with oceanographers at JPL. This mutually beneficial relationship helps the DAAC provide appropriate data, software, and tools, which in turn help the scientists conduct their research. A recent trend by JPL management to outsource

DAAC functions, however, could result in a loss of synergy with the scientific community and could ultimately undermine the oceanographic research objectives of the Earth Science Enterprise.

Although the DAAC serves the physical oceanography community well, it does not expend much effort to serve researchers in other disciplines. Some of these researchers will seek to combine disparate data sets from a variety of sources, including the PO.DAAC. Indeed, facilitating the creation of such multidisciplinary data sets is a primary goal of EOSDIS and was a reason for creating the DAACs and the ECS (see Chapter 1). Changing technology has reduced the necessity for the ECS, and as long as the DAAC is able to fully link its systems with the other DAACs, the goals of EOSDIS can likely be achieved without the adoption of the full ECS.

8

National Snow and Ice Data Center DAAC

Panel Membership

J.-BERNARD MINSTER, *Chair*, Scripps Institution of Oceanography, La Jolla, California
FERRIS WEBSTER, *Vice Chair*, University of Delaware, Lewes
VERA ALEXANDER, University of Alaska, Fairbanks
KEITH E. LITTLEFIELD, National Imagery and Mapping Agency, Fairfax, Virginia
JOHN E. WALSH, University of Illinois, Urbana
J. RONALD WILSON, Marine Environmental Data Service (retired), Manotick, Ontario

ABSTRACT

The National Snow and Ice Data Center (NSIDC) DAAC is hosted by the University of Colorado. Along with the NSIDC and the World Data Center (WDC) for Glaciology, with which it is inextricably intertwined, the DAAC manages data related to snow and ice, climate, and the cryosphere—the part of the Earth's surface that is perennially frozen. The NSIDC DAAC will also receive Moderate Resolution Imaging Spectroradiometer (MODIS) data from the GSFC DAAC and may create MODIS snow and ice products. Although no small task, particularly since the EOSDIS Core System (ECS) has not been completed, most of the EOS instruments related to the polar regions will be launched *after* the AM-1 platform. Consequently, the DAAC will face its most difficult data management challenges in a few years.

The NSIDC DAAC provides an outstanding example of how good data management practices and a close relationship with researchers can help lead to scientific advances. Although no major problems were found during its site visit, the panel recommends that the NSIDC DAAC sponsor joint activities with the ASF DAAC on scientific issues pertaining to polar regions, which have not received adequate attention from ESDIS so far. The panel also recommends that the DAAC develop and implement a transition plan describing the critical path of DAAC activities prior to site acceptance of the ECS.

INTRODUCTION

The National Snow and Ice Data Center DAAC was created by NASA in 1991 (Box 8.1). Its roots go back to 1957, when the World Data Center for Glaciology was established at the American Geographical Society in New York. The WDC relocated to the University of Colorado in 1976 with NOAA sponsorship, and a new data center, the NSIDC, was created in 1982. The NSIDC is by far the larger of the two organizations, and is funded by a variety of agencies, notably NOAA, NASA, and the NSF. The DAAC is larger still, and accounts for about 75% of the total operation. All three components are located within the Cooperative Institute for Research in Environmental Sciences (CIRES) at the University of Colorado.

All three components serve the cryospheric and polar science communities (Box 8.1). Although the operations and staff of the three centers are commingled, the holdings of the DAAC are distinct from those of the WDC and NSIDC. Current holdings of the DAAC include passive microwave and AVHRR products, altimetry and elevation data, and remotely sensed and *in situ* polar atmospheric science data.

The aggregate volume of these data sets, together with the holdings of the NSIDC and the WDC, is about 1 TB. In the EOS AM-1 era, the NSIDC DAAC will receive approximately 15-18 GB of MODIS data per day from the GSFC DAAC and will use them to produce MODIS snow and ice products. The DAAC will not be a direct recipient of high-volume Level 1 data streams until the Advanced Microwave Scanning Radiometer (AMSR) and the Geoscience Laser Altimeter instruments are launched in a few years (see Table 1.1 for a description of data processing levels).

To prepare for the upcoming missions, the DAAC is developing new products, testing MODIS algorithms, and working on the ECS release B testbed. All of its baseline hardware is in place, and the DAAC is reconfiguring hardware and installing the ECS commercial-off-the-shelf (COTS) software to prepare for Version 2. In terms of readiness for the EOS data streams, the greatest challenges that the DAAC faces are staffing up in time and planning the near-term transition to

BOX 8.1. Vital Statistics of the NSIDC DAAC

History. The NSIDC DAAC was created in 1991. Its operations are almost completely merged with those of the NSIDC and WDC for Glaciology, which has been disseminating data since 1957.

Host Institution. CIRES, University of Colorado in Boulder, Colorado.

Disciplines Served. Cryospheric and polar science; data are available on snow cover, freshwater ice, sea ice, glaciers, ice sheets, and ground ice.

Mission. To serve communities identified by the Mission to Planet Earth Strategic Enterprise Plan by providing easy and reliable access to EOS satellite data, ancillary *in situ* measurements, and any necessary baseline data, model results, and relevant algorithms relating to cryospheric and polar processes.

Holdings. The DAAC holds 1 TB of heritage data sets and anticipates receiving 15-18 GB of data per day from the AM-1 platform via the GSFC DAAC.

Users. There were 506 unique users in FY 1997, not including unregistered users who access the ftp site.

Staff. In FY 1998 the DAAC had 27 FTEs and 6 ECS contractors.

Budget. Approximately $4.1 million in FY 1998 (including DAAC costs and ECS-provided hardware, software, and personnel), increasing to $4.7 million in FY 2000.

the ECS. With regard to the first, the DAAC currently has funding for 33 FTEs, although several positions are vacant, and the DAAC will have to add 20 more FTEs over the next few years.

The Panel to Review the NSIDC DAAC held its site visit on March 4-5, 1998. The following report is based on the results of the site visit and on subsequent e-mail discussions with the DAAC manager in June through September 1998.

HOLDINGS

The snow and ice data archived and distributed by the DAAC (see Box 8.2) are a critical resource for the cryosphere research community. The need for

BOX 8.2. Data Holdings as of January 1998

- *Nimbus-7 Scanning Multichannel Microwave Radiometer (SMMR).*
- *Defense Meteorological Satellite Program (DMSP) Special Sensor Microwave/Imager (SSM/I).*
- *Advanced Very High-Resolution Radiometer (AVHRR)*—1-km Level 1B polar data set.
- *Sea Satellite (SeaSat) and Geodetic Satellite (Geosat)*—Altimetry data for the Antarctic and Greenland ice sheets.
- *Digital Synthetic Aperture Radar (SAR)*—Mosaic and elevation map of the Greenland ice sheet.
- *Historical Arctic Rawinsonde Archive*—Data from the early 1950s to mid-1996.
- *Arctic Water Vapor Characteristics from Rawinsondes*—Data from 1954 to 1991.

SOURCE: NASA (1998).

remote sensing data on snow and ice is likely to increase over the years as programs emerge addressing gaps in the predictive knowledge due to inadequate understanding of feedbacks related to the cryosphere. Large expanses of the Earth that have a permanent ice cover are remote and inaccessible, and remote sensing is the most, if not the only, effective tool for data gathering. This applies equally to glacial ice and sea ice. The need to detect changes over time will be a critical element. For example, the extent of snow cover in winter is an indicator of climatic conditions over the land areas in the northern hemisphere and is considered an index of warming conditions (e.g., Figure 8.1). A time series is therefore valuable for global change studies. Examples of other high-priority areas include the mass balance of glacial ice, which has implications for freshwater input into the marine environment, which in turn determines sea level rise and influences oceanic water column structure. Ice cover also affects surface albedo, modifying the energy balance of polar regions, thereby affecting global climate. These are only a few examples of why observations from space of snow and ice are critical.

Processing Plans

Most of the data sets currently held by the DAAC are received as Level 1 products, then processed into higher-level products. The panel notes that infor-

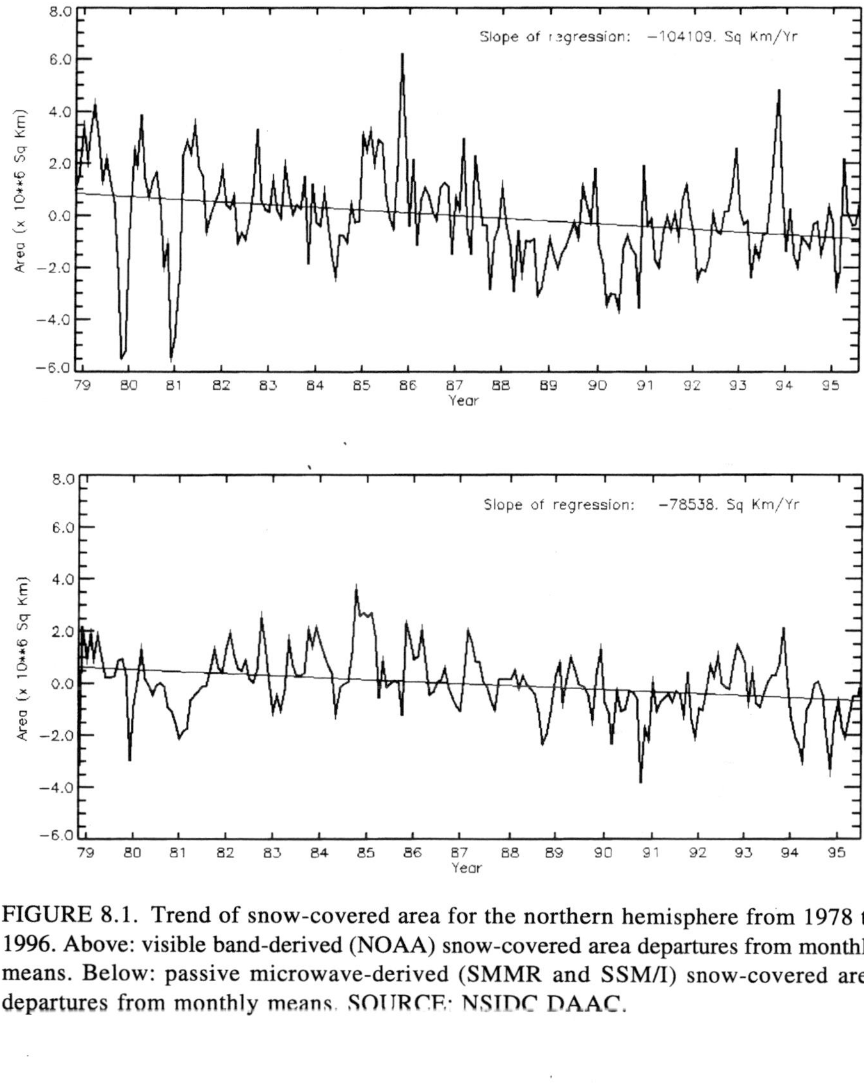

FIGURE 8.1. Trend of snow-covered area for the northern hemisphere from 1978 to 1996. Above: visible band-derived (NOAA) snow-covered area departures from monthly means. Below: passive microwave-derived (SMMR and SSM/I) snow-covered area departures from monthly means. SOURCE: NSIDC DAAC.

mation on lower-level versions of these data sets is not included in the DAAC's catalog. Some scientists need access to these data sets, sometimes in real time, and it is important for *all* the data holdings to be made visible in the catalog.

For the upcoming missions, NASA is considering transferring responsibility for data processing from the DAACs to the instrument teams on an instrument-by-instrument basis. The possible change in plans is being driven by new delays in the ECS. If the original processing plans are adhered to, the NSIDC DAAC will distribute and archive Level 1 AMSR data and will provide NSIDC glacier inventory information to validate ASTER data. It will also archive the derived Level 4 products from the ASTER Global Land Ice Monitoring System, although the ASTER images themselves will be made available by the EDC DAAC. Finally, the NSIDC DAAC will receive Level 2 MODIS snow and ice products from the GSFC DAAC and will create Level 3 gridded daily and composite products, which are not covered in the current MODIS instrument team processing proposals.

To prepare for MODIS product production, the DAAC has tested early versions of three out of four of the Product Generation Executables (PGEs). Drop 4 of the ECS software was delivered to the DAAC in April 1998, and the DAAC is currently testing the next version PGEs with the new software. None of the tests have used a simultaneous processing configuration, which is useful for determining whether the capacity of the system is sufficient.

Major Strategic Issues

Most users will not be able to manage the large files that will result from the EOS instruments. Subsetting will make the data more accessible to users and, in the DAAC's view, will probably be the biggest factor in increasing the size of the user community. Nevertheless, subsetting capabilities are not scheduled to be included in the ECS until 1999, *after* MODIS has been launched. Consequently, the DAAC cosponsored a workshop in July 1998 to discuss, among other things, the results of several prototype efforts for creating Hierarchical Data Format (HDF)-EOS subsetting tools. As a result of the workshop, the DAAC plans to test a subsetting tool developed at the University of Alabama, Huntsville.

Long-Term Archive

NASA and NOAA are currently negotiating plans to transition data from the DAACs to NOAA archives. The NSIDC has both funding and organizational links to NOAA, in the latter case, through a Cooperative Agreement between CIRES-University of Colorado and NOAA's National Geophysical Data Center (NGDC). Consequently, if NGDC becomes the designated archive, the transfer of data from the NSIDC DAAC to the NGDC should be one of the easiest to plan among all the DAACs.

USERS

Characterization of the User Community

The NSIDC DAAC tracks its customers by means of several criteria, including the number of users by category (e.g., university, federal employee, commercial sector), the number of data sets distributed, and the monthly volume of data downloaded from the Web. The NSIDC-WDC-DAAC complex also maintains an impressive list of several hundred journal papers than have used its data. The users of data and data products are multinational and diverse in discipline and technological sophistication.

High priorities of the DAAC include determining the scientific impact of data sets and ensuring customer satisfaction. A needs and requirements database that keeps track of user satisfaction is essential for the latter. Such a database extends beyond the DAAC's present tracking system, and the panel believes that the DAAC can and should be more proactive in its assessments of the uses of its data sets.

The panel suggests the following as possible ways to improve the assessments:

- identify and track electronic downloads of NSIDC DAAC data;
- determine users' satisfaction not only with the DAAC's provision of the data set, but also with the data set itself (e.g., What were the primary limitations the user encountered with the data set?); and
- assess the scientific impacts of the papers that resulted from the data set.

The last could be done by DAAC scientists and/or outside experts. This more extensive documentation would provide a basis for setting internal priorities on how much effort to place on particular data sets, and for general informational purposes for funding agencies, for example. By synthesizing information on data set utilization and impacts, the NSIDC DAAC can play a valuable role in evaluating the data priorities of the snow and ice community. As the voluminous and expensive products of EOS come on-line, there may be a much greater need than in the past for this type of function. The panel notes that the NSIDC DAAC is well positioned in this regard and would be missing a potentially important opportunity if it does not enhance its documentation and assessment of data set utilization.

User Working Group

The DAAC has a good relationship with its Polar DAAC User Working Group (PoDAG). The PoDAG meets every eight to nine months and helps the DAAC decide such issues as which nonstandard data products should be developed. The DAAC is responsive to PoDAG recommendations, and PoDAG members feel effective and productive. PoDAG members stated that the NSIDC DAAC is well integrated into the scientific community and is not simply a number-crunching data center. This characteristic arises in part from its cooperation with and proximity to scientific research programs (see below).

Because of their overlapping interests in the polar regions, the ASF and NSIDC DAACs had a joint User Working Group for the first few years of their existence. Apparently the joint group was not as effective as either the ASF or the NSIDC DAAC might have wished, and separate user working groups for the two DAACs were formed. The outcome, in the panel's view, was that the NSIDC DAAC gained an effective User Working Group, but at the cost of losing synergy with the ASF DAAC.

Recommendation 1. The NSIDC DAAC should sponsor joint activities with the ASF DAAC, such as joint meetings of the User Working Groups, on issues of mutual interest. In the panel's view, such an act of leadership would be beneficial to the polar sciences.

Interaction with the Scientific Community

The NSIDC-WDC-DAAC complex has a long and impressive history of responding to the needs of snow and ice researchers. Active involvement on the part of technical personnel in the acquisition and development of data products, and the close juxtaposition of the external support function with active faculty and student in-house research, have resulted in an understanding of the *modus operandi* of scientific research on the part of the technical staff and in a proactive attitude. The panel notes that this cooperative and proactive attitude is a strong positive attribute and that the in-house scientific competence adds value to the data sets. The cryospheric and polar science research communities will continue to increase their use of satellite remotely sensed data, as scientists who are not currently among the remote sensing specialists recognize the value of the data products and become users. Such an expansion of the number and diversity of users can well be accommodated by the DAAC.

Although the DAAC clearly has an excellent relationship with its scientific users, the review showed that visitors and outside collaborators play a relatively small role in the DAAC's operations. Among the reasons given were a lack of available space and computer equipment for visitors, as well as a tendency for in-house scientific staff to do the "value-added" work such as (1) synthesizing data sets, especially data sets of the same variable but from different time periods and/or regions; (2) reformatting and gridding data sets to facilitate user access; and (3) providing quality control and producing "clean" versions of data sets that were originally contaminated. The HARA Arctic upper-air sounding data set is an example of an NSIDC-enhanced data set that benefited a large segment of users. At present, the NSIDC tends to do this type of work when an in-house researcher needs the value-added data set. Hence, this work is often driven by external funding of in-house projects. This does not guarantee that the tasks undertaken are those most needed by the community. Visitors at the DAAC, who could contribute to such efforts, would bring a nice "external drive" into this process.

The DAAC would benefit in several ways from a more active visitor program. First, it would be able to tap into the scientific expertise and data set usage experience of the broader community, which is more diverse than even a high-quality in-house staff can possibly be. Feedback from external scientists working at the DAAC will almost certainly be more substantive than feedback from remote users. Second, a strong visitor program would foster the community's stake in the DAAC and would counter any perception of its being a "closed shop" or competitor. Finally, visitors would disseminate first-hand information about the DAAC, its holdings, and its products, thereby enhancing the visibility of the DAAC.

Possible vehicles for enhancing external collaboration include dedicated resources (e.g., workstations, space, travel funds) for visitors and a wide solicitation of visitors. In addition, collaborative ties could be fostered through joint research proposals by NSIDC DAAC personnel and outside scientists. The latter strategy represents a significant step beyond the practice of sending representatives to a workshop and submitting a proposal to serve as the data archive for a particular program (e.g., NSF's Arctic System Science initiative).

Recommendation 2. To broaden the scope of its interaction with the scientific community, the DAAC should sponsor a visiting scientist program with adequate space and should foster new collaborations with outside researchers.

User Services

The panel was impressed with the high level of user services that the DAAC provides to its customers. The balanced suite of analyzed products and data sets offered by the DAAC is much to the benefit of its clients. It also has an active CD-ROM publications program and services a significant number of regular subscribers. Subscribers to a series generally receive multiple products during a year. This is counted as a single request.

The method of counting requests is realistic in that the DAAC counts only requests for which it does work and supplies something to the user. It does not justify itself on soft figures such as hits on a Web site or requests that result in a simple referral to another center. Given the method of counting, the servicing of 1,500 to 2,000 requests a year (including more than 500 subscriptions) indicates that the DAAC is heavily used.

DAAC policy is that all requests receive a response within 24 hours. This does not mean that requests are always completed within 24 hours but that the request is acknowledged and the user receives information on how and when the data or information will be provided. This policy has contributed to a very positive relationship between the DAAC and its users.

Polar Grids

The ECS software employs rectangular projections for gridding data. Rectangular projections, however, are inappropriate for high-latitude regions because of the distortion in horizontal distance, which becomes increasingly severe toward the poles. At the pole, the calculations fail. Consequently, polar projections are essential to the cryospheric and polar science communities. The National Snow and Ice Data Center has developed the Equal Area Scalable Earth grid, which has been adopted by the community for the Polar Pathfinder data sets. However, if EOS data are not translated into polar grids, the DAAC (and NASA) will completely fail its user community.

At the time of the review, the DAAC had tried unsuccessfully for five years to make support of polar grids a requirement of the ECS. An instrument team had also submitted a request to the ESDIS Resource Allocation Board for funding to develop a polar grid. A subsequent discussion between the panel and the ECS contractor revealed that ECS contractors have begun to address the problem, and the panel urges ESDIS to ensure that this critical functionality is developed and fully tested before launch.

Recommendation 3. To ensure that the needs of the cryospheric and polar science communities are met, ESDIS should require that the capability to generate polar grids be incorporated into the ECS prior to launch of the AM-1 platform.

TECHNOLOGY

General Philosophy

The NSIDC DAAC seeks to balance the risks of prematurely adopting "cutting-edge" technology with the benefits that more modern systems provide. The DAAC therefore tries to be "sufficiently modern" with regard to hardware. The panel agreed that this is a sensible approach.

Hardware

The network topology and communication infrastructure seem sufficiently modern to meet the needs of the DAAC for the foreseeable future. The storage capacity of the StorageTek PowderHorn (300 TB) for on-line storage and archive is more than sufficient to handle the legacy data (1 TB) and the new data expected over the 10-year life cycle. The server and workstation configuration are currently adequate to meet the needs of the DAAC, although the number of workstations available for non-DAAC use (i.e., visiting scholars) was raised as an issue (see Recommendation 2).

Software

At the time of the site visit, the initial delivery of operational ECS software had not been made, and the DAAC was still working with Version 0. The panel was concerned about the configuration management of the software baseline and delivery process, given the number of sites and the mixture of core functionality and DAAC-specific functionality required. How will DAAC-unique functionality and "patches" be controlled, and how will they be integrated with future baseline upgrades and deliveries? The configuration management problem could be exacerbated if the DAAC runs Version 0 in parallel with Version 2 of the ECS. Maintaining multiple versions of the system software would also increase operations and maintenance costs.

Part of the problem is related to the lack of baselined standardized Application Program Interface (API) documentation, which was supposed to facilitate the development of DAAC-unique functionality and buffer the changes from the core functionality. To mitigate risk, the panel suggests that the ECS contractor work with the DAACs to set priorities for standardized API development and documentation, based on which functions are most critical, which will be used the most, and which are most feasible in terms of cost and schedule.

Processes

Operational testing of the baseline software in end-to-end functional tests is a critical step in determining whether the system is ready for production. Once this level of testing is completed (as part of the site acceptance tests), a detailed evaluation of the test reports, of outstanding hardware and software defects, of workarounds, and so forth, must follow as part of the assessment of operational readiness. At the time of the review, installation, checkout, and testing of the hardware were being performed by ECS contractors, and operational testing had not yet begun. In view of the timetable for the upcoming missions, the panel viewed this as an area of risk.

Media Versus Web Distribution Strategy

Although distribution of data on media will always be important, the DAAC regards Web distribution of its data sets and the development of home pages as major accomplishments. Over the past two to three years, ftp and Web access have increased sharply. The panel found the DAAC's Web site to be functional and useful, with many data sets that can be downloaded from the site.

MANAGEMENT

General Philosophy

The panel met with two managers at the site visit—Ron Weaver, manager of the NSIDC DAAC, and Roger Barry, director of the NSIDC and the WDC for Glaciology. Barry obtained funding in the early 1990s to create the DAAC and serves as the DAAC scientist. He also orchestrated a user community letter-writing campaign, which probably saved the DAAC from closure in 1994. The panel views Barry's involvement as a tremendous strength of the DAAC, not only because of his strong connections to the scientific community and his leadership abilities, but because he is a faculty member. The latter is particularly important because of the clout it brings with the host university.

The DAAC leverages considerable scientific and technical expertise through its association with the University of Colorado, its role as a World Data Center for Glaciology, and its many connections with international science programs such as the World Climate Research Program. These diverse and important associations ensure not only enhanced scientific and technical guidance, but also the broadest global view of the user community. This broad view of user needs is much to the benefit of the Earth Science Enterprise.

Operations Approach

The DAAC has adopted a "product team" approach to the development and management of its operations. A product team is responsible for a data set. The team is generally led by a scientist with expertise relevant to the data set, and other members are drawn from operations, user services, scientific programming, technical writing, and database administration. This approach helps ensure that all aspects of the life cycle of the data set are covered. To further guide the development and management of the data sets, data set production and documentation checklists have been implemented.

The panel considered this approach to operations and development of data sets to be effective. The lessons learned by staff working on one data set are transmitted immediately to the other product teams on which a member works. Methodologies are well established and understood. The procedures necessary to ensure that the breadth of expertise to build and manage end-to-end systems for complex, interrelated scientific data sets have been designed and formalized. Coupled with the checklists to ensure compliance, there is clearly a good management system in place to keep operations on track and on schedule.

Personnel

All DAAC staff, including the ECS contractors, are university employees. This further blurs the distinction between the centers and the university and between the DAAC and the ECS contractors. It also contributes to a strong sense of team membership. The staff have a mixture of expertise that is well suited to the mandate of the DAAC. In particular, there is strong scientific capability and leadership in the organization. This depth of scientific expertise has allowed the DAAC to develop a program and services that meet the real needs of its user base. In addition, the technical staff responsible for day-to-day operations, budgeting, and data set development were found to be knowledgeable, capable, and highly motivated. Considerable interest in the review was demonstrated by a large attendance at the open sessions with the panel. There was broad participation by all in the discussions, and contributions by the staff were relevant, positive, and to the point.

In general, the DAAC does not have a problem retaining staff. Turnover is about 8%, which is well below the rate of 30% in Colorado for all industries. It appears that the NSIDC complex offers an interesting and challenging working environment that results in the low turnover. Staff recruiting, on the other hand, appears to be more problematic. At the time of the review, there were several key vacancies, such as a system test engineer, at the DAAC. The system test engineer should have been in place before the hardware installation began. Filling this and other positions will help ensure that the DAAC is ready for the EOS data streams.

Budget

The NSIDC DAAC's FY 1998 budget is $4.1 million, about 10% of which is ECS-supplied hardware, software, and personnel (Table 8.1). About half of the resources are applied to development of new data sets and services, and about half to routine processing and operations. The panel views this as a healthy balance, particularly since science is moving toward multidisciplinary global activities that

TABLE 8.1. Total NSIDC DAAC Costs (million dollars)[a]

	Fiscal Year								
	1994	1995	1996	1997	1998	1999	2000	2001	2002
NSIDC DAAC	1.8	1.9	2.0	3.0	3.7	3.4	3.7	4.1	4.5
ECS hardware	0	0	0.3	4.6	0.1	0.7	0.4	0.2	0.1
ECS software	0	0	0	0.4	0.1	0.1	0.1	0.1	0.1
ECS personnel	0	0	0	0	0.2	0.5	0.5	0.5	0.6
Total cost	1.8	1.9	2.3	8.0	4.1	4.7	4.7	4.9	5.4

[a]Budget numbers for FY 1994 - 1997 are actual values; numbers for FY 1998 - 2002 are projections, as of May 1998.
SOURCE: ESDIS.

require the production of a broad suite of data and information products, rather than just the provision of data sets. It also allows the scientists who use the center to do science without having to do the data processing and analysis first.

In the panel's view, the budget and number of staff (33 FTEs) are reasonable, given the scope of responsibilities of the NSIDC DAAC. The cost of the DAAC and the split between salary and operations are in proportion compared to some other international data centers of similar size that deliver similar services. Leveraging off the activities of the NSIDC and WDC for Glaciology is also highly cost-effective, but the intermingling of these activities makes it difficult to determine the cost-effectiveness of the DAAC component by itself. In general, the DAAC would benefit from developing quantitative measures to show its cost-effectiveness, particularly given community concerns about the apparent high cost of the DAACs.

Contingency Plans

In the event of an ECS failure, the NSIDC DAAC has developed a contingency plan for MODIS that will cost a few million dollars a year for seven years. The MODIS instrument team has also developed a contingency plan, although details on the plan and its cost were not available to the panel. A final decision on which plan would be adopted had not yet been made. In either case, the DAAC risks failing its user community if sufficient funds are not available to produce the snow and ice products, or if implementing the contingency plans leads to a reduction in the number of geographic areas that can be processed—cryospheric science requires global coverage.

Strategic Plans

Future broad directions for the DAAC include delivering near-real-time data, creating a relationship with the modeling and climate assessment communities, and developing MODIS products. The first two were not discussed in detail at the site visit, and the panel assumes the DAAC has a detailed strategy for implementing them. The third relies, in part, on the technological transition from Version 0 to the ECS. The DAAC was concerned about setting priorities and managing competing demands for space and resources during this transition. It was not apparent in the review that realistic tests dealing with the anticipated large volumes of data and information, and with a possible large increase in the number of simultaneous user requests, have been attempted or that plans for such tests had been devised. The performance requirements developed by ESDIS, which are vague and will be difficult to verify, should be made more explicit. In addition, the panel suggests that as part of site acceptance testing, the DAAC run stress test procedures for average and peak usage to identify system bottlenecks.

Recommendation 4. The DAAC should develop a transition plan that describes the critical path of all DAAC activities that have to be completed prior to the start of site acceptance testing for future ECS deliveries.

NSIDC DAAC AND THE EARTH SCIENCE ENTERPRISE

Relation to CIRES

The director of CIRES, Susan Avery, recognizes that having the NSIDC-WDC-DAAC complex embedded in the university, yields benefits to both the university and the data center complex. In general, the university does not go out of its way to nurture the complex, but it does provide a reasonable level of infrastructure, including space, telephone and networking services, and contract administration. It also maintains the Internet connection to the DAAC. The University of Colorado's commitment to the DAAC will end when external funding ends.

The University of Colorado has provided university affiliation for DAAC technical personnel, and the administrative placement of the DAAC within CIRES has worked well. Although the intermeshing of the DAAC with its host institute creates complexities in financial and time budgeting, it leads to a high level of flexibility. The university environment and benefits are attractive to the staff, and the DAAC benefits in many ways from, and makes use of, its position within the university. In the panel's opinion, a tremendous strength of the NSIDC DAAC is the leadership of a faculty scientist who has a voice in university affairs. Consequently, the DAAC gains visibility and clout within the university.

Relation to NOAA

The National Snow and Ice Data Center is administered as a component of NOAA's National Geophysical Data Center complex of data centers and is located in a building adjacent to NGDC. Although NOAA does not have responsibility for the NSIDC DAAC, the proximity and administrative links between the centers present an opportunity for the DAAC and the Earth Science Enterprise. NGDC and the two other NOAA data centers (the National Oceanographic Data Center and the National Climatic Data Center) are the major archives of environmental *in situ* data in the United States. These data have enormous potential value to the EOS program, since they could and should serve to calibrate and validate satellite-derived data. NSIDC data will be used for ASTER validation purposes, but there is an opportunity for NGDC to use its links to NSIDC to play a larger positive role in the EOS mission. For that reason, the panel was disappointed that the connection between the NSIDC and the NGDC seemed *pro forma*.

Relation to ESDIS

The DAAC has had mixed results from its interactions with ESDIS. For example, ESDIS helped broker an agreement between the DAACs and the ECS contractor that will enable the DAAC to obtain the ECS source code before launch. On the other hand, it took several years for the DAAC to convince ESDIS of its need for polar grids. With the departure of Gregory Hunolt at ESDIS and Dixon Butler at NASA Headquarters, the DAAC is concerned that the idea of an integrated data system for science might be lost.

Relation to Other DAACs

The NSIDC DAAC interacts regularly with three other DAACs. It shares MODIS data product interdependencies with the GSFC and EDC DAACs, and it used to share the same User Working Group with the ASF DAAC, which also deals with polar region data. Most of the collaboration exists at the working level because the DAACs need to solve problems in common. At the management level, however, the relationship with the other DAACs is weak. Weaver told the panel that the DAAC managers worked together only because they had a common funding source and a strong manager at ESDIS (Hunolt).

Relation to the ECS Contractor

In the past, the DAAC communicated with the ECS contractor through ESDIS, an arrangement that led to frustrations on both sides. The DAAC is now communicating directly with the ECS contractor through its ECS liaison. The ECS science liaison came from the Raytheon ECS facility in Landover, Maryland, and has been able to open the lines of communication from the DAAC to the ECS contractor, although communication from Landover back to the DAAC is only beginning to be effective. Two-way communication will presumably help solve some of the frustrations that the DAAC has had with the ECS contractor, such as receiving incomplete and inconsistent documentation. It may also help reduce conflicts arising from the DAAC's desire to customize the ECS and the ECS contractor's desire to maintain a standard ECS configuration at all sites.

Relation to Instrument Teams

The DAAC is concerned about its limited ability to influence the interdisciplinary science teams and instrument teams. The DAAC staff believes it is essential to be involved with field programs if they are to do the data management well, but they have found that it is difficult to convince the scientists. Although six to eight DAAC staff are designated as mission data coordinators to the EOS instrument and interdisciplinary science teams and attend team meetings, the DAAC feels it does not have much influence over the data management plans.

SUMMARY

The NSIDC DAAC contributes well to the strategic goals of the Earth Science Enterprise by facilitating research that will lead to fundamental contributions to cryospheric and polar science. Two important factors have helped the DAAC implement its mission: (1) it has a strong vision of serving its science community; (2) it is collocated with two other major cryospheric-polar data centers, the NSIDC and the WDC for Glaciology. With regard to the first, the DAAC has an excellent relationship with its cryospheric and polar science user communitics. Indeed, the DAAC is embedded in the science community, an accomplishment that few data centers are able to achieve. Its success in this regard comes in part from understanding who its users are and what data they need. A better understanding of how the data are used, however, would further improve the DAAC's service to its customers. In addition, joint activities with the ASF DAAC would help the NSIDC DAAC develop closer ties with scientists who use synthetic aperture radar data in the polar regions.

The second factor for the DAAC's success, the collocation and intermingling of DAAC operations with the NSIDC and the WDC for Glaciology, not only leverages NASA's investment but also has practical benefits for the DAAC. The DAAC can take advantage of lessons learned by the older data centers, and it has access to a wide variety of ancillary data and in-house scientific expertise. Such in-house resources, however, come with the risk of becoming a closed shop, and the DAAC and its sister centers would benefit from a strong visiting scientist program.

The DAAC's relationship with its host institution, CIRES, is also good, in part because of the strong involvement of a faculty member, Roger Barry, in the DAAC. The DAAC's relationship with NOAA's National Geophysical Data Center, on the other hand, seems *pro forma* and the potential for a stronger, beneficial relationship has yet to be realized. A better relationship with NOAA could help smooth the transition for the long-term archive of NSIDC DAAC data.

One of the greatest challenges facing the DAAC is accommodating limitations in the ECS that will make it difficult for the DAAC to satisfy the needs of the polar science community. For example, the latest version of the ECS does not provide subsetting capabilities or polar grids, although the DAAC and the ECS contractor are taking steps to remedy these deficiencies. Although the DAAC appears to know what needs to be done to be ready for launch, a better near-term plan and schedule to transition from Version 0 to the ECS would help ensure the DAAC's readiness for the AM-1 platform.

9

Oak Ridge National Laboratory DAAC

Panel Membership

J.-BERNARD MINSTER, *Chair*, Scripps Institution of Oceanography, La Jolla, California
FERRIS WEBSTER, *Vice Chair*, University of Delaware, Lewes
KENNETH D. DAVIDSON, National Climatic Data Center, Asheville, North Carolina
BETH H. DRIVER, National Imagery and Mapping Agency, Chantilly, Virginia
WILLIAM J. PARTON, JR., Colorado State University, Fort Collins
ERIC T. SUNDQUIST, U.S. Geological Survey, Woods Hole, Massachusetts

ABSTRACT

The Oak Ridge National Laboratory (ORNL) DAAC manages data needed to study biogeochemical fluxes and processes. In contrast to other DAACs, the ORNL DAAC deals with data derived mainly from intensive field campaigns and process studies, rather than from satellites. ORNL DAAC data are important to the satellite program, however, because they provide a means for validating data from high-resolution imaging satellites, such as Landsat, and multisensor platforms, such as AM-1. In fact, without the integration of remote sensing and *in situ* observations, it is doubtful whether the EOS program would ever fulfill its full potential.

At the time of the site visit, the ORNL DAAC was only beginning to get involved in the EOS Land Validation Program. Given that Landsat 7 and the AM-1 platforms are scheduled to be launched less than a year from now, the DAAC will have to be resourceful and assertive to become meaningfully involved in the calibration and validation plans for these programs. It will also have to have a vision of what it intends to accomplish in the EOS Land Validation Program and other data activities, and the panel's main recommendation is that the DAAC develop a vision and implementation strategy for fulfilling its special mission within the DAAC system. A plan for participating in the EOS flight missions is a key part of the implementation strategy, and given the imminence of launch date, the DAAC needs to develop this plan immediately. Only by becoming involved with satellite programs will the ORNL DAAC effect the transition from being a biogeochemical data center to becoming a fully functional DAAC.

INTRODUCTION

The Oak Ridge National Laboratory DAAC was created in 1993 to archive and disseminate biogeochemical dynamics data from NASA and non-NASA field activities (Box 9.1). It is collocated with several environmental data centers at Oak Ridge National Laboratory, including the Carbon Dioxide Information Analysis Center, the World Data Center for Atmospheric Trace Gases, and the Atmospheric Radiation Measurement (ARM) archive.

The DAAC has a unique role within EOSDIS. Rather than serving as a repository for large volumes of data from a small number of remote sensing missions, it serves as collecting agent and repository for small, disparate sets of data from principal investigators of field projects sponsored by NASA and other federal agencies. Rather than processing data, the DAAC makes available data that have been processed by the scientists who collected them. These ground- and aircraft-based data are critical for validating EOS remote sensing instruments.

In the future, the DAAC will continue to provide data management support for field campaigns, such as the Large-Scale Biosphere Atmosphere Experiment in Amazonia (LBA), as well as for process studies related to biogeochemical dynamics. It will also place increasing emphasis on the EOS Land Validation Program. These activities will not pose a technological problem for the DAAC—processing capacities and response times are ample for current needs, and the DAAC is able to add on-line storage as required. In the panel's opinion, the greatest challenge facing the DAAC is implementing its role in integrating ground-based measurements into the overall array of remotely sensed data available from EOS and other NASA programs. The ORNL DAAC is a crucial link in the validation of biogeochemical inferences from remotely sensed data. It is not an exaggeration to assert that the success (or failure) of the ORNL DAAC in this role could make (or break) large components of the stated mission of the EOS

BOX 9.1. Vital Statistics of the ORNL DAAC

History. The ORNL DAAC, which was created in 1993, is one of several centers that manage environmental data in the Environmental Sciences Division of Oak Ridge National Laboratory. The DAAC's holdings go back to the 1800s, although most data sets are relatively modern.

Host Institution. DOE's Oak Ridge National Laboratory in Oak Ridge, Tennessee.

Disciplines Served. Biology, ecology, geology, and chemistry.

Mission. To provide data and information about the Earth's biogeochemical dynamics and ecology to the global change research community, policy makers, educators, and interested general public.

Holdings. The DAAC holds about 48,000 files with a total volume of 6-7 GB.

Users. There were 1,143 distinct users in 1997.

Staff. In FY 1998 the DAAC had 14 FTEs and 1 ECS contractor.

Budget. Approximately $2.6 million in FY 1998 (including DAAC costs and ECS-provided hardware, software, and personnel), decreasing to $2.4 million by FY 2000.

program. Yet, neither the DAAC, ESDIS, nor the instrument teams seem to appreciate fully the DAAC's critical role in EOS.

The Panel to Review the ORNL DAAC held its site visit on March 19-20, 1998. The following report is based on the results of the site visit and e-mail correspondence with the DAAC manager in June through September 1998.

HOLDINGS

The ORNL DAAC archives and distributes biogeochemistry data sets associated with intensive field campaigns and global terrestrial ecosystems process studies (Box 9.2). The intensive field campaigns, such as the First International Satellite Land Surface Climatology Project Field Experiment (FIFE), are funded through NASA's Terrestrial Ecology Program. The ORNL DAAC is preparing to receive the data from the Boreal Ecosystem-Atmosphere Study (BOREAS) field campaign and will manage the data from the LBA field campaign in the next few years. An example of data from a field site is shown in Figure 9.1. These data contribute to the overall goals of the ORNL DAAC by providing

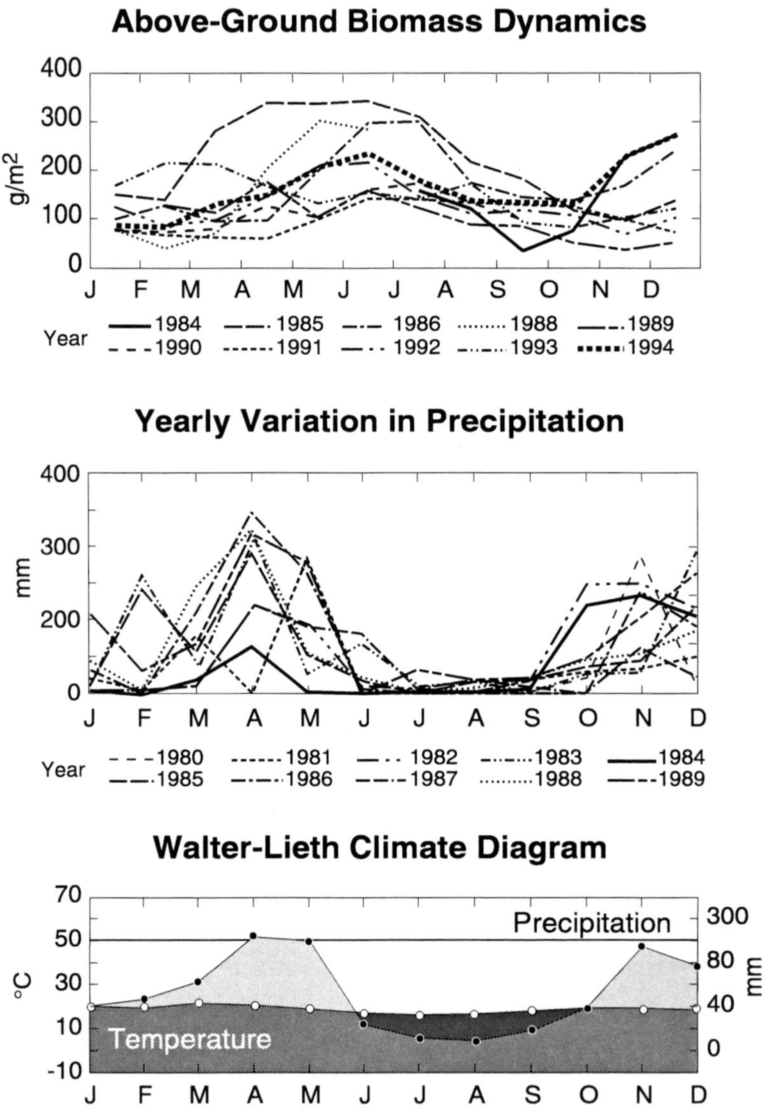

FIGURE 9.1. Response of plant biomass to monthly and yearly variations in seasonal precipitation at a grassland site in Kenya. The dry season, from June to October, is indicated by the Walter-Lieth Climate Diagram, a commonly used method of visualizing climate based on mean temperature and precipitation data (the dark shaded area represents the period of relative drought). Data on biomass response to climate variation are important for understanding the global carbon cycle and how net primary productivity on a grassland may vary in response to global change. SOURCE: ORNL DAAC.

BOX 9.2. Data Holdings as of January 1998

Intensive Field Campaigns

- *First International Satellite Land Surface Climatology Project Field Experiment (FIFE)*—Monthly data from the Kansas prairie for May 1987 to October 1989.
- *Superior National Forest (SNF)*—Daily, monthly, and yearly data from Minnesota for 1972 to 1990.
- *Oregon Transect Ecosystem Research (OTTER)*—Monthly data for May 1989 to June 1991.
- *Boreal Ecosystem-Atmosphere Study (BOREAS)*—Data are currently available from the Goddard Space Flight Center.

Process Studies

- *Net Primary Production (NPP)*—Daily, monthly, and yearly data from grassland and woodland sites worldwide. Studies range from 3 to 51 years in duration.
- *Amazon River Basin Precipitation*—0.2-degree gridded monthly and daily data from Peru, Bolivia, and Brazil for January 1972 to December 1992.
- *Global Wetlands and Methane Emissions*—Global monthly data at 1-degree resolution from the 1980s.
- *U.S. Streamflow*—Monthly data from the United States for 1874 to 1988.
- *River Discharge*—Measurements from more than 1,000 stations around the world for 1807 to 1991.
- *Hydroclimatology*—Monthly point data from the continental United States for 1948 to 1988.

SOURCE: NASA (1998).

data to test and develop biophysics and ecosystem models at different sites and to compare with biogeochemistry data that will be collected in the future. The global terrestrial ecosystem data sets (Box 9.2), such as Net Primary Production (NPP), are primarily used to develop and test global ecosystem models. The ORNL DAAC is also collecting and processing data from the Fluxnet program, an international program in which microclimate variables and fluxes of CO_2, trace gases, and heat are being measured from regional networks of eddy-correlation towers.

Commitment to Data

Perhaps the greatest strength of the ORNL DAAC lies in its commitment to data and its readiness to do the often thankless work required to render the data it receives useful to the larger scientific community. Such commitment is particularly important for data collected in terrestrial field studies in which investigators, often working in difficult and rapidly changing conditions, must make on-the-fly decisions to control the most significant variables in a unique setting. Rendering such descriptions useful to other scientists requires devoting time and attention to compiling, organizing, and presenting information about how the data were collected, and storing the data so that they can be readily used in conjunction with other data from the DAAC—work that the academic and research communities do not generally reward.

Integration of Ground-Based and Remotely Sensed Data

The ORNL DAAC's role in the EOS Land Validation Program will be (1) to make ground- and aircraft-based data available for validating satellite instruments and algorithms, and (2) to facilitate integration of these disparate data types by working with researchers to ensure that the data are in suitable, self-consistent formats. Both satellite data validation and data set integration are necessary for scientists to gain a more complete understanding of biogeochemical processes. With regard to validating the satellite measurements, one of the most important issues has to do with geolocation. For incoming satellite data and for modern field studies using the Global Positioning System (GPS), sources of data can be located on the Earth's surface precisely and accurately. However, it may not be possible to provide locations of comparable accuracy for many historical sources of data that are essential to understanding biogeochemical dynamics. Even modern field measurements must often change locations opportunistically, depending on local conditions, yielding aggregated data that may be representative of a less-than-precise location. Thus, for many biogeochemical studies, geolocation attributes must include estimates not only of position but also of geolocation precision and accuracy. Such estimates, which are often difficult and are seldom undertaken in a comprehensive manner, exemplify the kind of data treatment that should be provided by the ORNL DAAC. Yet, at the site visit, the ORNL DAAC did not seem to be aware of these issues.

Recommendation 1. As a crucial element of the EOS program, the DAAC should work to resolve the issues of accurate co-registration of *in situ* and remotely sensed data, and ensure that both its staff and its users fully understand what the DAAC's data holdings mean to the proper interpretation of remotely sensed data.

Metadata

Much of the DAAC's holdings are documented by the principal investigators of individual field campaigns and process studies, who tend to work with the data for three to four years until the quality control and documentation are as complete as possible. Incorporating metadata about field projects into the EOSDIS metadata model, which was designed for satellite data, is not direct and requires numerous interactions between the DAAC and ESDIS. In addition, the existing metadata descriptors are apparently changed by staff at the Global Change Master Directory (GCMD), often to descriptors that are not in common usage by field investigators, requiring the ORNL DAAC to expend additional effort to make such changes. For example, DAAC staff told the panel that ESDIS and GCMD staff have changed the descriptor "carbon dioxide" to "CO_2" to "atmospheric carbon dioxide concentration" and back to "carbon dioxide" over the past two years, requiring the DAAC to change its metadata files accordingly.

Recommendation 2. Given the importance of validating the EOS remote sensing measurements, ESDIS should ensure that the ECS metadata model accommodates data derived from ground- and aircraft-based studies.

Formats

Although the standard format for EOS data is HDF-EOS, all of the ORNL DAAC's holdings are kept as ASCII files. Users are happy with ASCII formats and ESDIS does not require that ground-based data be put into HDF-EOS because the overhead is too high for small data sets. An ability to work with HDF-EOS, however, will be needed for validating land data from MODIS and other remote sensing instruments, and the panel encourages the DAAC to become familiar with HDF-EOS.

Processing Strategy

Although the principal investigators of field experiments and process studies process most of the data eventually held by the DAAC, the DAAC acquires some unprocessed data from other sources. In addition to processing these data, the DAAC will become involved in processing Fluxnet data several years from now. On occasion, the scientific community has asked the DAAC to generate value-added products, but the DAAC has no immediate plans to reprocess data sets.

Long-Term Archive

NASA is currently negotiating with NOAA to provide a long-term archive for EOS data. Until a Memorandum of Understanding is concluded, the ORNL

DAAC will continue to archive its holdings as long as funding is available. The panel notes, however, that the ORNL DAAC is collocated with two long-term DOE archives—the Carbon Dioxide Information Analysis Center and the ARM archive. Moving the DAAC's data sets to one or both of those centers would likely be cheaper than moving them to NOAA and would keep the data sets where the scientific expertise resides. At the urging of the ORNL DAAC, NASA and DOE apparently considered an MOU on long-term archive several years ago. The panel urges the two agencies to resume their discussion.

Recommendation 3. NASA and DOE should consider establishing a Memorandum of Understanding for the long-term archive of biogeochemical data from the ORNL DAAC.

USERS

Characterization of the User Community

The ORNL DAAC primarily serves the global change research community, which includes scientists who use terrestrial ecology and biogeochemical dynamics data from process studies, field experiments, and remote sensing. A high proportion (about 30% of users in 1997) are foreign researchers, who learn about the data from scientific publications. Many U.S. scientists who are interested in the types of data held by the DAAC were originally involved in the field experiment or study and, consequently, are not frequent users. Individuals from private corporations (e.g., corporations that harvest forested lands) and institutions are also occasional users of the DAAC.

The DAAC does not keep track of the more detailed characteristics of its user community, and the panel encourages the DAAC to do so. This will permit the DAAC to develop a more accurate user profile, which in turn will assist it in expanding user activities and increasing usage of its data. The ORNL DAAC is arguably the most prominent U.S. data center focused on terrestrial ecology and environmental data, and it has the potential to serve a much larger segment of the biological and ecological communities.

In the panel's view, the DAAC should also begin to capture and more carefully analyze statistics on data usage and use the results to construct metrics of its performance. Not only would such data be useful in better understanding how users respond to DAAC initiatives or to external changes, but they would afford a deeper understanding of user differences and of gradual shifts in user behavior, which might be expected as holdings grow and become more diverse. A metrics program need not be elaborate; it must be inexpensive to implement, it must result in consistent data collection, and it must provide data that can be analyzed and used with ease and with confidence.

Recommendation 4. To better serve its users, the DAAC should develop and implement a strong metrics program to track resource usage, evaluate the impact of data management decisions, and predict the outcome of future actions.

Finally, the development of a "marketing" plan may also help the ORNL DAAC to expand its user base, thus increasing its constituency. This then could prove valuable for budget purposes as well as for generating successful proposals to other government or nongovernment agencies.

User Working Group

The DAAC has a good relationship with its User Working Group (UWG). The UWG is an independent body; it helps the DAAC determine which data sets to acquire and what emphasis to give to existing data sets. For example, the current allocation of work between field campaigns, validation of remote sensing products, and ecosystem modeling (see "*Data Priorities*," below) was recommended by the UWG.

The greatest concerns of the UWG include (1) increasing the size of the DAAC's user community; (2) competition for DAAC activities by the Earth Science Information Partners (ESIPs) of NASA's prototype federation; and (3) transferring the BOREAS data, which are currently being managed at Goddard Space Flight Center, to the DAAC so that they can be distributed to the broader community. With regard to the latter, the BOREAS Project Office and the ORNL DAAC formally agreed several years ago that the BOREAS data would be transferred to the ORNL DAAC for archive and unrestricted distribution. However, neither the DAAC nor the UWG has been able to obtain the data, with the result that the transition is taking a longer time than expected.

Relationship with the Scientific Community

A strength of the ORNL DAAC is its willingness to work with the scientific user community to provide information needed for research. The DAAC is working toward involving scientists more closely in its operations, and an obvious opportunity for enhanced interaction exists at Oak Ridge National Laboratory, which hosts a large environmental sciences group. The DAAC could also enhance its visiting scientist program or recruit DOE postdocs to work with the data. Seeing scientists work with the data will help the DAAC understand better how the data are used and how it can better serve the needs of the scientific community.

Past experience with scientists who provide data to the DAAC has shown the DAAC the benefits of early involvement in the intensive field campaigns. If the DAAC has not been involved in any stage of the experiment, it will have a considerably more difficult time serving users when the data finally arrive. The

DAAC has taken important strides to remedy this problem through its early involvement in planning data management for the LBA intensive field site investigation, and the panel applauds this strategy.

User Services

The panel got the impression that the user services group at the DAAC is dedicated and competent. Indeed, the DAAC prides itself on its ability to satisfy users requests. If the DAAC is to meet the challenge of the EOS Land Validation Program, however, the user services group will have to place increased emphasis on providing standardized data sets for the development and intercomparison of biogeochemical models, and on providing data from a variety of platforms (chambers, buoys, towers, shops, aircraft, balloons, satellites) in self-consistent and accessible ways.

TECHNOLOGY

Hardware Availability

The DAAC's in-house computer suite appears to be a reliable resource for DAAC staff and users. Processing capacities and response times are ample for current needs, and the project is able to add on-line storage as it is required. The team appears to use adequate configuration management practices and to perform operations housekeeping, such as backups, consistently. One user observed that network capacity for on-line delivery of data to customers was likely to become a problem; however, it is expected that most DAAC users will continue to use public networks, augmented by transportable media.

The panel believes that developing a simple performance model for end-to-end servicing of customer requests would help the DAAC evaluate the net impact of response times at any point in the process. For example, the impact of high-speed retrieval capabilities for data to be delivered by Federal Express would have to be weighed against other services the DAAC might improve with the same funds.

Significant changes in system load may occur when large volumes of time-series data are collected and delivered to customers on a regular basis. The DAAC will have to tune performance as resource scarcities shift, to anticipate sudden degradations in service as resources saturate, and to justify budgeting for increased capacity while current resources are not fully utilized. The outyear budgets presented to the review team did not show significant funds earmarked for additional system capacity.

The DAAC is not on the schedule for ECS delivery and will soon have to replace its hardware. It has no plans for an evaluation of costs and benefits of using ECS as opposed to continuing to use a relatively small, tailored system with

ECS-compatible data definitions and interfaces. The DAAC should be prepared to make a case for its findings on either side of this issue.

Finally, the panel suggests that within the next four to six months, the DAAC verify its own readiness for the millennium rollover. Although the DAAC team was able to answer specific questions concerning potential year-2000 problems and appears to have avoided a significant data conversion cost, it does not appear to have done the methodical self-assessment and verification called for by U.S. government directives, which are warranted to ensure uninterrupted service.

User Interfaces

Developers showed justifiable pride in fielding easy-to-use user interfaces and evaluating them based on user behaviors. They also showed creativity in applying new technology to expedite the work of in-house staff as well as to facilitate timely, complete, and accurate capture of relevant data from scientists at the point of capture. Such tools, combined with prompt processing of the data upon receipt, should reduce costs and improve the fidelity of the descriptive data that accompany field samples.

Media Versus Web Distribution Strategy

Some DAAC staff voiced a preference for Web technology (i.e., string-based searches or browsers that can be used with many file structures) over traditional database management system technology. The panel counsels that such a change should not be made before more powerful Boolean search capabilities become available. As the DAAC's holdings grow, users may have greater need for the precise search capabilities afforded by structured databases than they do at present. Web browsers often return large numbers of irrelevant references that are difficult to eliminate with today's search tools.

MANAGEMENT

General Philosophy

The DAAC management team, led by Larry Voorhees, is well experienced in data management activities, and this is reflected in its ability to instill into the staff the requirements and importance of data management. In general, the basic data center functions of acquisition, quality control, archiving, and providing access to data appear to be well understood by the DAAC and carried out successfully. Several CD-ROMs based on current data holdings have been produced, and they demonstrate the DAAC's understanding of the value of and requirements for its data holdings by the research community. The strengths of the management activities are tied to the staff's dedication to the data, not the organization, and the general understanding of data center activities, which is due in part to their collocation with other Oak Ridge National Laboratory data centers.

In the panel's view, DAACs should be more than data centers because, in addition to the data center functions mentioned above, they are involved in active satellite programs. The ORNL DAAC, however, functions more like a data center than a DAAC. To become a DAAC in reality, the ORNL DAAC will need a clear identity and a stronger sense of its special mission (validating satellite data with ground- and aircraft-based measurements) within the EOS program. A vision and an implementation strategy for achieving this vision are critical. The implementation strategy should also describe the DAAC's participation in the EOS flight missions. Such participation in the flight missions will require a proactive attitude at the DAAC. In fact, the DAAC's experience with early involvement in upcoming field campaigns such as LBA will help it form similar relationships with the instrument teams. By thinking strategically, the DAAC will also be able to further improve the effectiveness of its data center operations.

Recommendation 5. The ORNL DAAC should articulate a vision of its mission within the EOS program and an implementation strategy with goals for fulfilling this mission. The strategy should influence every decision the DAAC makes, from participation in the relevant EOS flight missions, to targets for data acquisition, priorities for data ingest and preparation, staffing, development of a user base and constituency, and allocation of resources for current-year and projected spending. Such a strategy should help the ORNL DAAC become a DAAC in reality.

Personnel

The DAAC has 1 ECS liaison and funding for 14 FTEs, which supports 10 full-time and 13 part-time staff. Only a few FTEs are devoted to system development, indicating the DAAC's focus on operations. DAAC staff come from a variety of line organizations at Oak Ridge National Laboratory, but they are located together and report to Voorhees. The line managers supervise the staff, write annual performance reviews, and make decisions on promotions with input from Voorhees. Consequently, the DAAC has little opportunity to directly reward or promote outstanding efforts.

Turnover appears to be higher than normal, perhaps because of uncertainties in long-term funding. Employees seem to recognize that their jobs may, and could, go away at the discretion of a NASA executive.

The DAAC has a good relationship with its ECS liaison. She is viewed as a team member by the DAAC, and her only ECS tasks are to keep track of the status of ECS deliveries to other DAACs and to make sure that the DAAC does not stray too far from the ECS standards.

In the panel's view, the ideal data center has a mixture of data management experts, computer programmers, and scientists familiar with the types of data stored at the center. Scientists familiar with the data have a first-hand understanding of the potential problems associated with experimental data sets, the scientific value of the data sets, and the type of analysis being done. The excellent work by Jonathan Scurlock (ORNL scientist) and Dick Olson (DAAC staff) on the global NPP data sets shows the advantage of having scientists familiar with the NPP data work with data management personnel. In general, however, scientists in the Environmental Science Division do not appear to be actively working with DAAC personnel on most of the biogeochemistry data stored by the DAAC, and the DAAC should aggressively pursue opportunities for collaboration (see "*Relationship with Scientific Community*").

Budget

The DAAC's FY 1998 budget is approximately $2.6 million, and the budget is not expected to vary greatly over the next several years (Table 9.1). Staff costs are currently about 85% of the total DAAC budget, indicating the DAAC's focus on operations, rather than development. The DAAC practices full-cost accounting, so the figures given in Table 9.1 represent the true cost of the DAAC, including occupation of the facility and support services.

Although the budget is small compared with the other DAACs, the UWG is concerned that the DAAC's user community is too small to justify the DAAC's resources. Based on the annual budget for the program and the number of staff supported, it appears that the average cost per work year is somewhat high, even when the academic credentials of the staff are considered. The costs are driven by

TABLE 9.1. Total ORNL DAAC Costs (million dollars)[a]

	Fiscal Year								
	1994	1995	1996	1997	1998	1999	2000	2001	2002
ORNL DAAC	0.8	2.7	2.1	2.2	2.6	2.3	2.3	2.5	2.7
ECS hardware	0	0	0	0	0	0.4	0	0	0
ECS software	0	0	0	0	0	0.1	0	0	0
ECS personnel	0	0.1	0.1	0.1	0	0	0.1	0.1	0.1
Total cost	0.8	2.8	2.2	2.3	2.6	2.8	2.4	2.6	2.8

[a]Budget numbers for FY 1994 - 1997 are actual values; numbers for FY 1998 - 2002 are projections, as of May 1998.
SOURCE: ESDIS.

the overhead imposed by ORNL operations. If the DAAC grows in order to process more data sets, it should consider organizing its work so additional staff would not require the academic credentials and experience typical of the current labor mix.

The panel also suggests that the DAAC capture actual costs in a work breakdown structure that would permit isolation of costs for specific services performed. It would be useful to capture other performance indicators in order to analyze the high "cost per stored byte" that is the only available performance measure at present and to evaluate impacts of variables, such as source or type of data, and process changes, such as those for facilitating data capture that the staff has already initiated. The panel recognizes that the acquisition, quality control, and maintenance of the ORNL data sets (each of which is unique and requires unique management) is labor intensive and suggests that these costs be quantified as part of the performance measures. The fact that the DAAC practices full-cost accounting provides a rare opportunity to estimate the true cost of operating a data center. The absence of hidden costs may partly explain the apparently high cost per stored byte.

Data Priorities

The ORNL DAAC archives and distributes biogeochemistry data sets associated with three major activities: (1) NASA intensive field campaigns, (2) EOS Land Validation Program, and (3) global terrestrial ecosystems process studies. Currently 40% of the DAAC's effort is associated with archiving and distributing the small, diverse data sets from the intensive field campaigns; and another 40% of the DAAC's effort goes toward the EOS Land Validation Program. The remaining 20% of the DAAC effort is concerned with global terrestrial ecosystem data sets.

The 40:40:20 allocation of effort was recommended by the DAAC's User Working Group, and it represents a significant increase in emphasis on the EOS Land Validation Program. The panel concurs with this change and suggests that in the future the DAAC should further increase its emphasis on the EOS Land Validation Program, if it can do so without jeopardizing its ability to meet the data needs of the broader biogeochemical community. Archiving and distributing data from NASA-funded field campaigns contribute to the overall objectives of the ORNL DAAC but should not be its primary focus. By expanding its role in the EOS Land Validation Program, the ORNL DAAC will strengthen its position in the overall EOS program.

ORNL DAAC AND THE EARTH SCIENCE ENTERPRISE

Relation to Oak Ridge National Laboratory

The ORNL DAAC is located within the Environmental Sciences Division of Oak Ridge National Laboratory. The Environmental Sciences Division focuses on five strategic areas, three of which are relevant to the DAAC—response of ecosystems to environmental change, integrated assessments, and environmental data management. The division's budget is $40 million per year, which is divided among about 75 activities.

To the panel, the DAAC appeared to be lost in the Oak Ridge organization. If NASA funding is reduced, there seems to be little motivation for the Oak Ridge general contractor to perform this type of work in the future. Similarly, the DOE office at the laboratory views the ORNL DAAC as "work for others," which will cease to exist when funding ends.

Relation to ESDIS

The ORNL DAAC feels that ESDIS listens to its needs and concerns, but it is a small voice in a big system. Its main advocate within the NASA management structure is Diane Wickland, the DAAC's program manager at NASA Headquarters, who provides advice and guidance to the DAAC. The panel felt that without Wickland's influence, the DAAC's position with ESDIS would be precarious at best. Although the DAAC's main interaction with ESDIS is through William North, the DAAC felt it had a good relationship with Gregory Hunolt, former DAAC system manager. At the time of the review, Hunolt's departure had led to considerable uncertainty about the DAAC's future interactions with ESDIS.

As noted above, the ORNL DAAC is the smallest of the DAACs, but it has the crucial role of validating EOS instruments. Therefore, in the panel's view, it should have an equal voice at ESDIS, and appropriate consideration should be given to meeting the special needs of the DAAC, such as incorporating field-related key words into the ECS.

Recommendation 6. ESDIS should devote greater attention to the importance of the ORNL DAAC to the success of the EOS program, support its activities as a full player in EOSDIS, and thereby help it become better integrated within the DAAC system.

Relation to Other DAACs

Although the ORNL DAAC works with the EROS DAAC on the EOS Land Validation Program, it apparently has little other interaction with the EOSDIS system. The DAAC perceives the weekly telephone conferences with other

DAAC managers as focusing on ECS issues, which are not relevant to the ORNL DAAC, rather than on problems that DAACs face in common. It participates only because it is politically important to do so. The DAAC has great difficulty working within the EOSDIS system and would operate differently if it had a choice.

Relation to ECS Contractor

In the past, the DAAC had a difficult time working with the ECS contractor because ECS activities are contract driven, and many of the DAAC's requests to ECS are not included in the contract. For example, the DAAC has had a major problem with the ECS contractor in getting key words accepted in the metadata model.

The ORNL DAAC has never been scheduled to receive the entire ECS because it does not manage satellite data. In the last year, the DAAC was removed from the delivery schedule (although it may receive selected components in the year 2000), so there is no longer a need to hold technical discussions with the ECS contractor.

Relation to Instrument Teams

Until now, the "instrument teams" for the ORNL DAAC were the principal investigators of the NASA intensive field campaigns and process studies. The DAAC has generally had a good relationship with these data providers and has positioned itself to become more involved in the planning of future studies, such as LBA and Fluxnet. The panel urges the DAAC to become similarly involved with the EOS instrument teams that need data from the ORNL DAAC to validate their instruments and algorithms. Given that MODIS is scheduled to be launched in a matter of months, it is imperative that working relationships between the ORNL DAAC and MODIS instrument and science teams be developed immediately.

SUMMARY

The ORNL DAAC has two important roles in NASA's Earth Science Enterprise—to facilitate *in situ* ecological science research and to validate remote sensing data from the EOS satellites. The latter—involvement in an active satellite program—is what distinguishes a DAAC from a data center. With regard to its data center functions, the ORNL DAAC is arguably the primary U.S. data center focused on ecological data. It has done pioneering work on the acquisition and distribution of biogeochemical data sets, and the careful stewardship of its holdings is impressive. Indeed, one of the greatest strengths of the DAAC lies in its commitment to data. The DAAC makes good use of its User Working Group to identify data sets to acquire, but it should also take advantage of the scientific expertise that resides at Oak Ridge National Laboratory and/or strengthen its visiting scientist program. A closer relationship with scientists would help the DAAC understand more about its holdings and how they are used.

DAAC staff have also proven to be creative and innovative in providing tools that users need. To satisfy the evolving needs of its users, however, the DAAC must embark on near- and midterm strategic planning. A vision and an implementation plan for fulfilling this vision must be developed for the DAAC to fully succeed in its data center mission.

The need for a vision and implementation plan is even greater in the DAAC's other mission in the Earth Science Enterprise—validation and calibration of EOS satellites. Unless *in situ* data from the ORNL DAAC are integrated successfully with the satellite data, the interpretation of satellite observations cannot be validated. Yet, plans for resolving the geolocation problem inherent in integrating these disparate data types have not been developed or even conceived, and the ECS metadata model does not accommodate key words needed to search and retrieve land- and aircraft-based data. Further, neither the DAAC, ESDIS, nor the instrument teams seem aware of the importance of involving the DAAC in the planning stages of the flight missions. Early involvement of the DAAC would help ensure that the *in situ* data needed for validation and calibration are available in a form that is suitable to the instrument teams. This issue must be resolved immediately because the space missions (e.g., Landsat 7 and the AM-1 platform) will be launched within a year. To get involved at this late date, the DAAC will have to become more aggressive with ESDIS and the instrument teams. It can no longer wait passively if it wishes to fulfill its special mission within the EOS program. Fortunately, the DAAC can draw on the experience of its successful involvement in the planning stages of the LBA intensive field campaign.

Afterword

This report describes EOSDIS as it was configured between January 1997 and August 1998. Since that time, NASA has sponsored two important meetings that directly affect the DAACs: (1) a workshop on plans for the long-term archive of EOS data in particular, and earth science data in general; and (2) a meeting on data management strategies for the Earth Science Enterprise beyond the year 2000. The archive workshop, held in November, was organized by the U.S. Global Change Program Office, NASA, and NOAA. It resulted in a set of draft recommendations on functional specifications for a long-term archive, including topics such as metadata and assuring data quality. These specifications will form the basis for determining the cost of the archive project and are expected to be incorporated into a proposal within NOAA for an FY 2001 initiative to archive DAAC and other large data sets. This was perhaps the first interagency workshop ever held on this topic, and the Committee is gratified that U.S. science agencies are taking steps to resolve these issues.

NASA's first meeting on post-2000 data management strategies (dubbed the "new DISS" initiative) was held in October. The DIS initiative builds on the federation concept and PI-led data management. It was driven by delays in the ECS, the recognition that smaller, more flexible systems can manage data more efficiently now, and the need to inject more science into data management. NASA's current plans are to retain the DAAC-ECS-ESDIS system for managing data from the AM-1 and Landsat 7 missions, although backup systems, rather than the ECS, will be used to process much of the data. (The ECS contractor will focus on delivering hardware and software to support users). When the new DISS model is adopted, the data will be managed by DAACs and/or data management services, depending on whether NASA decides to recompete the DAACs. In either case, peer review, such as that which has been offered so effectively to NASA by the space science community, will be essential input to deciding which data nodes should be continued or closed if they do not perform well.

References

Asrar, G. and R. Greenstone, eds., 1995, *MTPE/EOS Reference Handbook*. National Aeronautics and Space Administration, NP-215, Washington, D.C., 276 pp.

Independent External Review Panel (NASA), 1997, Assessment of 1997 MTPE Biennial Review Letter to Daniel Goldin, 6 pp.

NASA, 1988, *Earth System Science, A Program for Global Change*. Report of the Earth System Sciences Committee, National Aeronautics and Space Administration, Washington, D.C., 208 pp + supplemental brochures.

NASA, 1998, *Catalog of Data Sets and Services Available from the EOS Distributed Active Archive Centers (DAACs)*. National Aeronautics and Space Administration, Washington, D.C., 57 pp.

NRC, 1986, *Issues and Recommendations Associated with Distributed Computation and Data Management Systems for the Space Sciences*. National Academy Press, Washington, D.C., 111 pp.

NRC, 1994, *Panel to Review EOSDIS Plans; Final Report*. National Academy Press, Washington, D.C., 88 pp.

NRC, 1995a, *A Review of the U.S. Global Change Research Program and NASA's Mission to Planet Earth/Earth Observing System*. National Academy Press, Washington, D.C., 96 pp.

NRC, 1995b, *Earth Observations from Space: History, Promise, Reality*. National Academy Press, Washington, D.C., 310 pp.

NRC, 1996, Final Conclusions on the U.S. Global Change Research Program and NASA's Mission to Planet Earth/Earth Observing System. Letter to Robert Corell, 18 pp.

NRC, 1998a, *Global Environmental Change: Research Pathways to the Next Decade*. National Academy Press, Washington, D.C., in press.

NRC, 1998b, *Toward an Earth Science Enterprise Federation: Results from a Workshop*. National Academy Press, Washington, D.C., 51 pp.

APPENDIXES

APPENDIX A

DAAC Review Agenda

First Day

8:00	**CLOSED SESSION:** Overview of site visit plan and goals	*Minster*
	OPEN SESSION: DAAC REVIEW (EVERYONE WELCOME)	
8:30	Welcome and introductions	*DAAC manager*
8:40	Opening remarks	*Minster*
8:45	Mission of the DAAC • What is special about this DAAC, and why do we need it? • What are the special considerations for the data and information holdings? • How does this DAAC fit in with the others? • What steps is this DAAC taking to make its holdings usable together with those from other DAACs?	*DAAC manager*
10:30	Break	
10:45	Response to Criteria for Review	
	1. Relationship to MTPE Mission	*DAAC personnel*
	2. Data/Holdings	*DAAC personnel*
	3. Users	*DAAC personnel*

12:00	Working Lunch (NRC review team meets with chair of User Working Group)	
1:00	Criteria for Review (continued)	
	4. Technology and Facilities	*DAAC personnel*
	5. Management	*DAAC managers*
2:30	Break	
2:45	One-on-one visits	
4:00	**CLOSED SESSION:** NRC review team discussion	
5:30	Meeting adjourns for the day	

Second Day

8:00	**CLOSED SESSION**: NRC bias and conflict-of-interest discussion	*Linn*
9:00	**OPEN SESSION**: Revisit major issues with DAAC personnel	*Minster*
10:30	Break	
10:45	Wrap-up session	
	• Summary	*Minster*
	• Outstanding issues	*Minster*
	• Questions of the review team	*DAAC personnel*
11:30	**CLOSED SESSION AND WORKING LUNCH**	
	• Initial conclusions and recommendations	
	• Writing and homework assignments	
3:00	Meeting adjourns	

APPENDIX B

Criteria for Review

The following criteria for review provide a checklist of issues for the DAACs and the NRC site visit panels. Not all of them are equally important to all the DAACs, but they are included to ensure that all relevant topics have been covered. Following the review criteria are some suggested measures of performance. These measures are examples of the type of evidence that the DAACs may choose to bring forward to show their success in meeting the review criteria.

Relationship to MTPE Mission

- Maintaining or enhancing position as a primary source of data and information for the global change research community
- Being ready for data streams from EOS and other flight missions
- Establishing and maintaining a working relationship with providers of data products, algorithms, and ancillary information
- Fitting within MTPE strategic plan

Data/Holdings

- Providing access to data, data products, and information of high integrity and quality (including issuing data updates, notifying users of errors, and facilitating access to data from international sources)
- Fostering the quality of holdings (EOS and non-EOS data, produced by instrument teams or otherwise)

- Ensuring that data and data products are properly documented and that all appropriate ancillary information is available
- Ensuring that data and associated metadata are secure (e.g., data and associated metadata are not lost, destroyed, or insufficiently backed up)

Users

- Characterizing the user community (e.g., Who are they?) and setting priorities among different user community needs. Are the various user communities satisfied?
- Ensuring that holdings keep pace with research needs
- "Marketing" and advertising holdings
- Educating and reaching out to users about the application and significance of data and information holdings
- Fostering a sense of ownership among the users (e.g., promoting active participation by the users in DAAC activities)
- Cooperating with other agencies and international partners to ensure availability of science data to the widest possible user community

Technology and Facilities

- Keeping up with technological advances
- Using appropriate technology within the constraints of the EOSDIS requirements. Is it sufficiently modern? Is it cost-effective?
- Ability of the DAAC to choose the hardware and software technology it uses within the constraints of the EOSDIS requirements. Is there a plan for what equipment to acquire?
- Knowledge and use of appropriate national and international data standards

Management

- Responding to user feedback and emerging data and information needs
- Entraining, hiring, and retaining high-quality personnel
- Implementing effective mechanisms for setting and revising internal priorities (e.g., data handling, processing, backlog absorption)
- Promoting local innovation and initiative
- Ensuring that an adequate level of quality control is maintained in an operational setting
- Participating in an effective process for resolving different priorities among EOS units (federal and nonfederal)
- Interacting and cooperating with other DAACs to address common issues and avoid duplication of effort

- Developing adequate and practical working relationships among responsible agencies and international partners, where applicable
- Leveraging personnel, technology, scientific guidance, et cetera, from a host data center or other scientific institution, where applicable
- Developing and implementing a five-year plan for the DAAC (personnel, computers, etc.)
- Given the extent and quality of data holdings, satisfaction of the users, appropriateness of the technology and facilities, skill of the management, and so forth, is the DAAC cost-effective?

SUGGESTED MEASURES OF PERFORMANCE

The DAACs should have the primary responsibility for demonstrating that they meet the above review criteria. The following measures of performance are some suggestions that may be helpful:

- Statistics of processing user requests (e.g., turnaround time, track record in filling orders, error rates)
- Profiles of users by categories, such as which users dominate the total use; to what types of institutions do occasional users belong; are users seeking or providing information; are they instrument team scientists, other global change scientists, or the general public; are they domestic or foreign users; are they occasional or frequent users, etc.
- Number of publications involving use of data or information provided by the DAAC
- Short list of the most important scientific advances in which the DAAC has had a significant involvement
- Number of users and growth in number and diversity of data requests
- Spirit of initiative (e.g., DAAC innovations affecting global change research)
- Regular interaction with the scientific community (important for quality control, relevance of holdings, setting priorities, and keeping pace with new research needs)
- Leverage DAAC activities with resources from parent or other institutions
- Turnover of personnel
- Independence of the User Working Group, the degree to which it has input into the agenda for its meetings, and the degree to which management is responsive to the Working Group recommendations

APPENDIX C

CGED Interview with ESDIS (September 1997)

PART I. What does ESDIS expect of the DAACS? ***[questions based mainly on the EOSDIS DAAC Strategic/Management Plan, hereafter referred to as Plan]***

1. Relationship with scientific user community.

Science advisory groups have been established to ensure that the data and information services provided by the DAACs meet the needs of the science community (Plan, section 2.4). How well are the DAAC advisory groups working? To what extent is their advice coordinated with advice from other scientific advisory groups?

In general, the DAAC User Working Groups (UWG) have been very effective in providing DAACs with guidance as to their individual strategic plans and priorities for functionality and content. These priorities are particularly helpful with establishing holdings of the DAACs, as well as in determining needed community outreach.

This effectiveness is illustrated by the efforts of the DAACs to respond to the recommendations that come out of the UWG meetings and the willingness of the UWG to make recommendations to NASA on behalf of the DAACs. In particular, the DAACs are required to obtain their UWG's concurrence on their annual work plans prior to submission. These plans are used by the ESDIS Project as input to the negotiation of budgets and work priorities.

In addition, UWG membership has been fairly stable, despite the investment in time required to participate in regular meetings. We conclude that the members feel that this activity is a reasonable investment of their time, either for ensuring that their own needs are met or that the wider communities' needs are met. With respect to the latter, UWGs make particular efforts to ensure that the group's composition represents a cross-section of the potential user community.

The other advisory groups advise as to EOSDIS capabilities, and, therefore, indirectly the DAACs, as the long-term institutional base for the EOSDIS. The coordination between these groups is mainly via NASA, with additional informal coordination by way of dual participation in the meetings of these advisory groups. Several of the DAAC UWG members are also members of the EOSDIS Data Panel which is responsible for advising the ESDIS Project in overall direction and priorities in serving the science community. In addition, DAAC Project scientists and DAAC Managers participate directly in the Data Panel meetings.

DAACs are kept informed of any advice Project receives from other science advisory groups such as ESSAAC [Earth System Science and Applications Advisory Committee], NRC Panels, ERG [EOSDIS Review Group].

2. System development.

Each DAAC is responsible for determining whether extensions to existing ECS system capabilities are necessary (Plan, section 3.1.2). Have the DAACs responded to ESDIS with proposals for system improvements or prototyping? Which of these have been incorporated into EOSDIS?

DAACs have suggested and been funded to perform system improvements and prototypes via both their annual work plans and more formal proposals to the ESDIS Prototyping Manager. The latter proposals are more directed at solutions for evolving ECS (and are reviewed by the EOSDIS Data Panel), while the former have been more directed towards building and improving V0 [Version 0].

Two "official" prototyping activities are currently in an early implementation test phase at the GSFC DAAC. They are 1) U. of Maryland dynamic query preview capability in the V0 IMS [Information Management System] browse interface and 2) Georgia Tech web-accessible case-based reasoning help tool based initially on V0 user requests to the DAAC help desk. The query preview capability was also incorporated into the ECS JEST tool.

ECS has funded prototyping activities with universities, including UAH [University of Alabama, Huntsville] (partnered with the GHRC [Global Hydrology Resource Center] at MSFC [Marshall Space Flight Center] has developed and operated the Hydrology DAAC) to implement subsetting which was incorporated into their products.

DAACs have participated in the various reviews of ECS development and in EOSDIS working groups addressing the design and operation of the ECS. Discussions of extensions (especially additional data products) have started in some UWG meetings, and it is expected that these plans will become better defined once the ECS is operational, the DAACs have experience in working with ECS and its interfaces, and standard products are regularly generated.

3. System integration and testing.

Hughes has missed several deadlines for releasing new versions of the EOSDIS software. All of the DAACs expressed concern about these delays. With the available ECS software, are the DAACs ready to receive, process, disseminate, and archive EOS data? If not, what are the likely impacts on specific DAACs and instrument teams?

The initial delivery of ECS software to the DAACs (the Pre-Release B Testbed) is supporting the initial testing of science software within the DAAC environment. The August demo version of ECS will be installed at the GSFC DAAC in late October to support early operator training and science software testing in single thread scenarios. A test version of the system will be deployed at GSFC and EDC in January 1998 and at LaRC and NSIDC in February 1998 to support science software integration and testing in the production environment and to support critical interface testing. This will support preparations for operational readiness in mid-May (GSFC, EDC) and mid-June (LaRC, NSIDC). Following the launch of AM-1 at the end of June and Landsat-7 in early July and the subsequent instrument turn-on period, the DAACs will be ready to receive, process, distribute, and archive data.

To mitigate any risk of further delays in ECS, teams of ITs [instrument teams] and DAACs have been developing Emergency Back-up Systems to support minimal production and distribution of products for 6 months following the launches of AM-1 and Landsat-7.

The EOSDIS Alternative Implementation Path has also been studied with inputs from ITs (and DAACs) as a contingency against a failure to deliver the full ECS,

4. Operations.

During the V0 timeframe, the DAACs are responsible for all local operations, including data reception, product generation, cataloging, archiving, distribution, and reprocessing, as well as defining staffing and resource requirements (Plan, section 3.5.2). How well V0 is performing? How much of the anticipated demand can be met by V0? What lessons learned from Version 0 have been incorporated in EOSDIS?

Version 0 has been very successful in providing data to the users, as is illustrated by the metrics in our monthly reports. However, V0 has minimal real-time ingest (ASF the exception) and product generation requirements. The LaRC and GSFC systems are being augmented to support CERES processing for TRMM, but a major development would be required to provide an automated production environment sufficient to support MODIS and to provide the production on demand functions required for ASTER. With the Version 0 systems, the DAACs are currently managing a volume of data comparable to only 6 months of AM-1 and Landsat-7 data. A major expansion would therefore be required in the data management systems and archive capacity. Furthermore, while fully distributed, the Version 0 systems do not provide an evolvable architecture. Development would be required at several of the DAACs to support PM-1 and follow-on missions.

At the start of the ECS contract, a Version 0 "lessons learned" paper was developed by the Project and the DAACs and used to guide the ECS planning. The DAAC's experience in Version 0 has also guided ECS through their on-going involvement in design reviews and working groups. For example, much of the Version 2 metadata model and user interface look-and-feel is derived directly from the corresponding V0 items and community feedback on these items, including the on-going V0 Web user interface project. Some of the most important lessons to be incorporated from the V0 effort into Version 2 work have been processes for soliciting and incorporating community feedback into Version 2 development. This has been particularly important with regards to working metadata and user interface issues.

The DAACs generate products and/or distribute products that were generated by the instrument teams. What successes and problems have been associated with these activities?

DAACs have successfully worked with PIs [principal investigators] in defining and producing various Pathfinder products, products from existing flight missions such as ERBE and SAGE. NSIDC's work with EASE-Grid has been highly praised by the user community.

The DAACs are working with the Instrument Teams in the integration and testing of science software into the ECS environment. This collaboration is working very well. There has been one instance where a key ECS member was not available at a DAAC during an Instrument Team visit and the communications needed to obtain backup support did not follow the planned procedure, but the process has been corrected to avoid a similar problem.

How well are the DAACs documenting, reporting, and resolving operational problems within a DAAC or between DAACs?

Weekly telecons of the DAAC Managers and quarterly face-to-face meetings, facilitated by the ESDIS, give opportunity for reporting and resolving problems, as well as a mechanism for addressing common problems or assisting each other. For example, recently NSIDC found that it could provide the GHRC (previously operating the Hydrologic Cycle DAAC) much needed spare optical platters. DAACs worked together to smoothly transition data sets into their various archives when it became necessary to close the Hydrologic Cycle DAAC, while continuing cooperation with the GHRC (the hosts of the Hydrologic Cycle DAAC). Other on-going cooperative efforts allow DAACs to obtain data streams needed to meet their requirements (for example, SSM/I data from MSFC to NSIDC, other products between EDC and NSIDC, and between GSFC and LaRC). No major problems have bubbled up to the attention of the ESDIS Project.

For both ECS IR-1 delivery (January, 1996) and ECS Pre-Release B Testbed (April, 1997), a conflict resolution system has been in place to report, document, track and facilitate the resolution of problems and issues found in the activated ECS, at each DAAC. Once documented, a committee comprised of representatives from ESDIS, HAIS [Hughes

Applied Information Systems Corporation], and each DAAC meet, via telecon, to clarify new problem reports, determine the status of outstanding problems, and determine strategies to resolve the more difficult problems.

In addition, after Instrument Teams (ITs) and DAACs completed their Science Software Integration and Test (SSI&T) operational activities on ECS IR-1, each DAAC/Instrument Team contributed to the ESDIS produced document entitled: ECS IR-1 SSI&T Lessons Learned. This document is an excellent account on how to avoid unforeseen problems when integrating science software into the ECS. SSI&T is an ECS activity that will continue as long as science software is improved and/or science analysis evolves. A call for lessons learned from SSI&T into the ECS Pre-Release B Testbed has been made, and inputs are currently being gathered from the ITs and DAACs

The Zraket report noted that the DAACs have little if any budgetary control in the ECS. What budgetary and staffing discretion does ESDIS allow the DAAC managers?

Separate budgets are established for each DAAC for meeting high-level requirements, and each DAAC, within the guidance of its UWG, has discretion in determining how to most effectively meet those requirements. In addition, DAACs often successfully propose additional work over and above the budgetary guidelines. The DAACs have worked closely with the ESDIS Project in developing operations staffing and transition plans whereby a significant portion of the budget originally assigned to ECS has been transitioned to the DAACs for operation of the ECS-delivered system at the DAACs.

The DAACs have opportunity to provide input to ESDIS's evaluations of the ECS Contractor's performance, and a DAAC representative sits on the ECS Contractor's Performance Evaluation Board (for fee determination).

5. User support.

Given the fragmented responsibility between the DAACs and the instrument teams, how are the DAACs expected to provide user support on technical issues related to data quality and product generation?

The responsibility of the instrument teams is to develop algorithms and software for generating the standard products, and perform operational quality assessment on the data products as they are produced. Instrument teams may request the DAACs to help in the quality assessment, but the ultimate responsibility for data quality rests with the ITs. User support from DAACs are for both "push" users (i.e., the instrument teams) and "pull" users (consumers of data).

In support of the "push" users, DAACs support the Science Software Integration and Test process both initially and for any subsequent updates. DAACs and their respective ITs have worked out the SSI&T procedures and identified their respective responsibilities. The ESDIS project has facilitated this process through SSI&T workshops. ECS Contractor provides any support needed in the SSI&T process since the science software has to work in the software environment developed and delivered to the DAACs by the contractor.

To support the pull users, DAACs are expected to be aware of the products they hold, their characteristics, utility, anomalies, etc. We recognize there is a learning process involved here and the Instrument Teams are expected to provide some initial help in getting the DAAC personnel educated about the specifics of their products. The ITs are also responsible to provide the appropriate documents (ATBDs [Algorithm Theoretical Basis Documents], Guides, etc.) to the DAACs for storage and distribution along with the products. During the initial months of a product's existence, there will be need for consultative assistance from the ITs to the DAAC user services personnel. This is part of the scientists' responsibility in publishing their data. Providing this assistance in the beginning to the DAACs will relieve them of the long-term burden of supporting user inquiries.

6. Science software development, integration, and test.

The science software produced by EOS investigators is integrated into the DAAC production environment, where it is tested to verify its readiness for operational product generation (Plan, section 3.7). How much software integration and testing has taken place?

Each DAAC has been an integral part of, and is in concurrence with, all discussions involving Science Software Integration and Test (SSI&T) schedules, plans, and instruments and algorithms. Although the goal is to perform SSI&T on all science software before launch, it is essential that science software to perform these high priority functions, be integrated and tested with ECS.

The largest effects that ECS delivery delays have had on science software is in the area of better understanding the ECS/Science Software interfaces. Delayed ECS leads to delayed SSI&T, and thus a delay in understanding of these interfaces. The final result is a delay or postponement of implementing enhancements to the interface and Science Software. Sometimes lessons learned from one delivery are barely known in time for the next delivery.

Fortunately, the primary interface between ECS and the Science Software is the Science Data Processing Toolkit which has not suffered any delays. Delays in the readiness of ECS has not affected the science software readiness schedule.

To what extent are the science teams behind on their software deliveries? How has the readiness schedule been effected by the delays in delivery of the ECS software?

[Note: this question was not answered.]

7. Science data planning.

To what extent has the DAAC experience with user interests and satisfaction affected NASA's planning for future product generation and product support services?

The DAAC experience has led NASA to fund the DAACs to perform various services, including a CD-ROM sampler which can be useful to a large segment of the user community and prepackaged data sets which save users effort. In addition, the DAAC experience has emphasized the needs for establishing common formats to help the users, as well as lower costs of providing services.

PART II. What do the DAACs expect of ESDIS? *[questions based mainly on advance team visits to the DAACs]*

1. Relationship with scientific user community.

The EOSDIS Project Scientist (Skip Reber, formerly Steve Wharton) brings science issues related to the DAACs to the attention of the relevant science advisory groups (Plan, section 2.4). How well does this process resolve conflicts in priorities or gaps in communication between the DAACs and the science producers?

A Data System Working Group (DSWG) chaired by the EOSDIS Project Scientist facilitates communication and action among the instrument teams, DAACs, and the ESDIS Project. Members of the DSWG include representatives from the ITs, DAACs and the ESDIS Project. Representatives from the "pull" user community have been included in the meetings as well when relevant. This group is especially important to resolving priorities when there are budgetary limitations, as well as resolving data product needs. For example, when the ECS deliveries were delayed and development activities had to be replanned, this working group was called on to assist in developing a list of functional priorities for ECS Releases B.0 and B. 1. Also, a recent meeting was held of the DSWG to resolve metadata issues and answer several questions that the ITs have had regarding their responsibilities in providing metadata. (Due to the approaching launches of Landsat-7, AM-1 and SAGE-III, a high priority has been placed on getting the system ready for data producers and addressing issues related to generation of standard products.)

2. *System development.*

The Zraket report noted that the DAACs have little control over the management of the ECS or its future evolution. Indeed, some DAACs said to the CGED advance teams that the ESDIS response to their questions and comments takes so long that it is not possible to contribute meaningfully to ECS development. One DAAC noted that the DAACs and ESDIS worked collectively to implement Version 0, but that there was little opportunity to contribute to the development of Version 2. How do you respond?

ESDIS has followed a different paradigm for the development of Versions 1 and 2 from that for Version 0. (Last December, the responsibility for Version 1 development to support the TRMM launch was shifted from the ECS contractor to the Goddard and Langley DAACs due to schedule slips by the contractor and to help focus the contractor's attention on Version 2.) For Versions 1 and 2, the system development was not subdivided into parts that DAACs were independently responsible for and a part that was to be developed collaboratively. The rationale for this was the need for a large common core system infrastructure—thus the EOSDIS Core System (ECS). It was felt that duplication of the extensive functionality needed to handle the required automation, throughput, etc. would be too expensive with distributed development. This approach ensured that the DAACs were involved in the development of the ECS by involving them in all the ECS design and requirements reviews (several DAAC personnel served on each review board) and workshops (most DAACs had several personnel attend each of five user interface workshops), various design/operations working groups, provided liaisons (science and system engineering) to the DAACs from the ECS contractor, and provided additional funds to each of the DAACs to support up to 3 FTEs for interacting with the ECS contractor. Through their participation on review boards, the DAACs have made many suggestions for changes to system functionality (e.g., ability for end users to retrieve parts of science software packages), implementation (e.g., placement of browse data on higher performance archive) and architecture (e.g., location of product request handling in the system) that have been incorporated into ECS. Recently, ESDIS also made the DAACs part of the newly formed Integrated Test Program, wherein the DAACs will play an integral role throughout the ECS test lifecycle, not just operations.

In replanning the ECS Release B development, the DAACs were involved (along with the ITs) in setting the priorities, reviewing the detailed functionality allocated incremental releases, in setting the success criteria for the August demonstration, and in reviewing the August demonstration. There has been a growing need on the part of the DAACs to gain a better understanding of how ECS will function in order to plan their operations. The August demonstration provided this perspective and has elevated this planning to a new level.

3. System integration and testing.

The DAACs are expected to participate in system tests, assess the results, and recommend the acceptance of any deliverables (Plan, section 3.2.2). However, several DAACs feel that ECS is a moving target with large uncertainties in what will be delivered to the DAACs and when. How can the DAACs test a system for which there is no complete picture and which changes constantly?

A Test Methodology Working Group, which includes ESDIS Project personnel from the DAAC and Science Operations addresses this issue. It is defining the inputs to testing (such as system documentation) and the types of test that must be performed by various groups. The DAAC managers are kept abreast of the status of this group and solicited for their input through the weekly DAAC Manager's Telecon and the weekly DAAC Operations Working Group.

4. System management.

The ESDIS Project is responsible for establishing user service, data products, and processing requirements for the DAACs (Plan, section 3.3.1). How much weight should the DAACs place on providing user services to the general public?

DAACs are guided by the priorities set by the "EOS Execution Phase Project Plan," a contract between NASA Headquarters and GSFC. This document provides the "Level 1 Requirements" for all the EOS-related Projects including EOSDIS. The pertinent requirement here is:

> The EOSDIS shall provide Earth science data and information services to the EOS investigator community, the broader Earth science community, policy makers, the education community and the general public in that order, commensurate with resources.

However, it is expected that under the current budget DAACs can support the general public. This does not mean that DAACs are expected to spend a substantial proportion of their budgets to support this segment of the user population. In fact, DAACs have discovered that some small-scale activities can meet many of the needs of this segment. However, DAACs are not funded to duplicate or compete with the other outreach activities conducted by NASA, or other government agencies. If a DAAC has ideas for a substantial activity it is advised to propose to these other groups (such as NASA's education outreach group) or through their own institutions (for example, University of Colorado at Boulder).

Recently, the EOSDIS Review Group sponsored by NASA HQ reviewed the high-level goals of EOSDIS and revised this requirement as follows (in their draft report):

(C) By processing, archiving, and distributing environmental data, EOSDIS will support the following user communities:

- a) National and international agencies and entities with whom NASA has written agreements or legal obligations concerning MTPE data
- b) NASA-funded MTPE investigators
- c) Members of the broader U.S. and international Earth science community
- d) U.S. policy makers

If levels of support are sufficient, the following communities will be served to the extent possible:

e) The U.S. education community
f) The general U.S. public
g) Others

While this change in requirements has not been formalized yet, the service to the general public is at a lower priority than that to the other categories of users in both the present and revised versions of the requirements. Also, NASA has a commitment to education, and some level of support to the U.S. education community will be provided, even if resources are not sufficient to cover categories (a-d).

The ESDIS Project is also responsible for overseeing the hardware system-wide, but some DAACs are concerned about having the appropriate technology to support the transition from Version 1 to Version 2. How do you respond?

Technology selection is based on, among other things, buying hardware as late as possible to take advantage of price/performance improvements, maximizing processing power and storage capacity to meet requirements, facilitating porting of instrument team software from science computing facilities, and supporting software compatibility from version to version within ECS.

5. Operations.

Several of the DAACs noted that the attention of ESDIS and Hughes has been strongly on systems development, rather than on operations or maintenance of the system. Will the transition to an operational mode with more focus on the DAACs occur in a timely and smooth manner? Are sufficient funds available to support long-term system maintenance?

The transition to operations is not occurring at the most optimum time because of the development delays. However, through the Test Methodology Working Group mentioned above, we are attempting to ensure that operations needs are met as best as possible given the other circumstances.

In addition, we have found it difficult to justify our operations and maintenance budgets. This means that budgets continue to erode, as in many areas, but in particular this area. This is a problem in that there is not functionality that can easily be cut to solve the budget problems. We are going to have to do a better job at justifying these budgets. However, it may take actual cost data, as opposed to our models to convince the decision-makers of this need.

The ECS contractor assigns some staff to the DAACs, but the level of contractor support can be very unstable. Why not allow the DAACs to do their own recruiting?

The ECS Contractor was originally requested to provide staffing for all DAACs with major operational responsibility for EOS instrument data, with expectations that operations would be transitioned to the DAACs at a later date. This set the stage for the operations concept for the system to be developed: How many people would be required, at what skill, etc. Recently DAACs were given the opportunity to move up this transition, and three DAACs have taken advantage of this opportunity: NSIDC, JPL, and LaRC. (The JPL and NSIDC plans were approved by the ESDIS Project a couple of months ago; the LaRC plan was approved over a year ago.) EDC (except for user services) and GSFC have declined to take on this responsibility until later in the operations phase. And even the DAACs that took advantage of this opportunity requested that certain functionality continue to be provided by the ECS Contractor.

In addition, the ECS Contractor has been working with DAACs to ensure that personnel recruited work well in the DAAC environment.

6. User support.

The DAACs, in consultation with their User Working Groups, are responsible for evaluating user needs (Plan, section 3.6.2), but at least one DAAC said that ESDIS inserted different priorities, thus inhibiting interaction between the DAACs and their users. How do you respond?

NASA may sometimes have to direct DAACs to support additional requirements for programmatic reasons, such as commitments to other agencies or foreign partners. If additional funds do not come with these requirements, then ESDIS and the DAACs must accommodate the requirements by diverting resources from items of low priority.

Such decisions should be made with the cognizance of the DAAC's UWG. The SMP [Strategic/Management Plan] addresses this possibility in a section on "Developing Priorities for Data Set Support."

7. Science software development, integration, and test.

The attention of the instrument teams and EOS project has been strongly on algorithm development, rather than on smooth and consistent operations. At least one DAAC noted that frequent changes in algorithms are inconsistent with the production of readily interpretable time series. How will an appropriate balance be assured when the system becomes operational?

> The priorities in the early stages of the mission (several months after launch) will be on instrument and algorithm checkout. Therefore, there is a need for frequent changes. However, when the algorithms are more stable and it is scientifically meaningful to produce standard products routinely, the operations will be more consistent and smoother. In "steady state" operational setting, it is at the SCFs [Science Computing Facilities] that the frequent updates to algorithms will be implemented and tested, and it is after they are certified that the software migrates to the DAACs. The ECS delivered to the DAACs is sized to accommodate operations in parallel with I&T [integration and test] of new versions of algorithms. Decisions to reprocess data on a large scale will be made by the Data Processing Resources Board chaired by the EOSDIS Project Scientist (Skip Reber) and with representation from the science community including the ITs.

Some DAACs thought that the instrument teams were reluctant to provide documentation on the algorithms and quality control methodologies used in processing data products, making it difficult for the DAACs to serve their users. How will ESDIS ensure that the instrument teams provide the necessary documentation to the DAACs?

> The ITs are required by their contracts to provide Algorithm Theoretical Basis Documents (ATBDs). In addition, through the DSWG meetings, the ITs have been persuaded to provide simplified Guide documents about their instruments and products. Through a QA Working Group set up under the auspices of the DSWG, the Project has been working with the ITs and DAACs to ensure that the ITs provide their QA plans and make their requirements on the Project and the DAACs known, and identify and resolve any related problems.

8. Contingency planning.

Several of the DAACs feel that the ECS is overdesigned (i.e., one-stop shopping) and may even be an obstacle that the DAACs have to "wire around." Given the delays in ECS development, could the DAACs work together to evolve a unified system?

It would be difficult for the DAACs to develop such a system in support of AM-1 or Landsat-7. DAACs can somewhat scale their current systems to address some particular needs, as the LaRC and GSFC DAACs have to support TRMM instrument data while allowing the ECS Contractor to concentrate on the bigger problems of AM-1 and Landsat. However, they would have difficulty meeting the larger processing and data requirements of AM-1, Landsat-7, and future missions with the various V0 systems, without the infrastructure to be established with the ECS.

The ECS has been designed with the premise that originated from the science community's advice that one-stop shopping (i.e., consistent searching and ordering/accessing data at multiple locations) was a desirable thing to do in support of interdisciplinary science. Requirements have been discussed with the various advisory groups in order to determine whether any of the requirements could be relaxed. The EOSDIS Review Group has suggested some potential modifications, however no group has recommended dropping the one-stop shopping requirement. This requirement is particularly valuable for the larger user community which does not currently have access to large amounts of NASA data. (Also, while one-stop shopping is often cited as a function that could be reduced to save a significant amount of money, it is not a cost driver.)

When the ECS development is completed, it will provide a scaleable, adaptable, and evolvable architecture that is easily extended to support new missions and data products. It will be highly automated, support user access to information without requiring knowledge of data center holdings, and permit DAAC-unique extensions to support individual science discipline needs.

If time permitted the development of independent systems by the DAACs to meet AM-1 and Landsat-7 data loads, the DAACs could evolve an interoperability layer as was done under Project leadership in Version 0. However, ECS provides a lower level of interoperability, for example, support for product generation fed by data product subsets from other DAACs.

As a contingency, the instrument teams are constructing their own, home-grown processing systems that are independent of EOSDIS. What steps will be taken to involve the DAACs in this development?

Strictly speaking, the word EOSDIS in this question should be replaced with ECS. The ESDIS Project has funded each of the ITs to develop Emergency Back-up Systems without depending on any of the yet-to-be-delivered ECS capabilities. Each IT was asked to propose an appropriate teaming arrangement with DAACs to ensure that, through the period of launch through 6 months thereafter, the IT/DAAC team could produce, archive and distribute the minimal set of data products necessary to checkout, calibrate and validate their instruments and algorithms. (Note that performance requirements are greatly relaxed for these backup systems.) It was left to the ITs to determine the best teaming arrangements in each case. The following is the set of teaming arrangements that has resulted:

ASTER: Data processing at JPL SCF; Archival and storage at EDC DAAC [L1 processing occurs in Japan, data are shipped to and archived at EDC; EDC sends small subset of L1 data to JPL SCF per JPL IT's request; JPL SCF produces higher level products and transfers to EDC DAAC for archival and distribution]

CERES: Data processing, archival and distribution occurs at Langley DAAC using the "Langley TRMM Information System (LaTIS)" appropriately augmented to provide additional capacity needed for early months of AM-1 data processing.

MISR: Processing at JPL SCF; archival and distribution at Langley DAAC.

MODIS: Processing at MODIS TLCF at GSFC; archival and distribution through GSFC, EDC and NSIDC DAACs for the appropriate products

MOPITT: Processing at NCAR SCF; archival and distribution at Langley DAAC.

PART III Other issues *[questions based mainly on the Zraket report]*

1. Several key management positions are vacant at NASA HQ and Goddard. For example, some of the DAACs feel that the departure of Dixon Butler left a vacuum, and all are concerned that Charlie Kennel has not been replaced. This issue was also raised in the Zraket report. How do these vacancies affect the long-term stability of EOSDIS?

Dixon Butler certainly left a vacuum. Some DAAC UWGs are concerned that, although they were originally chartered by NASA HQ, NASA HQ no longer gives them any guidance. Dixon's vision, initiative, and leadership helped develop and communicate a common purpose. His programmatic responsibilities now reside in the Science Division at Headquarters and in the MTPE Program Office at Goddard. However, the science leadership for developing the objectives for EOSDIS is now very ably provided by Skip Reber, the EOSDIS Project Scientist. Skip has taken an active role, for example, in establishing processes for setting priorities and allocating EOSDIS resources to science products.

The long term stability of EOSDIS will depend on the success of the system in satisfying science user needs and on user acceptance of some of the compromises necessary to achieve a broader good (e.g., the acceptance of metadata and format standards and system capabilities needed to make data usable by a broader community).

2. *Who has end-to-end responsibility for making sure that the relationship between the science (people and knowledge) and EOSDIS is going smoothly?*

Skip Reber, the Acting EOSDIS Project Scientist is responsible for this. H. Ramapriyan, Chief of the ESDIS Science Office provides Skip Reber with the necessary support from the Project. The ESDIS Project Manager, Rick Obenschain, is responsible for assuring that EOSDIS meets science requirements within budget and schedule commitments.

3. *What do the EOS investigators (instrument teams and interdisciplinary investigators) think of their DAACs?*

From the comments we have heard at the Investigators' Working Group meetings and other science advisory groups' meetings, the DAACs and the services they have been providing are being viewed quite favorably. This is also evidenced by the demand for the DAACs' services from the user community. Further evidence is the intense resistance we saw from the community when a change to a "Hub and User Services Centers" concept was proposed by the Project about two years ago as a cost-saving measure.

Similarly, UWGs include EOS investigators, and in general the UWGs are very happy with their DAACs, as evidenced by various letters of recommendation received by the ESDIS Project and MTPE. The investigators' willingness to continue to be associated with the UWG, and the required time commitments, imply that the investigators think this is a worthwhile effort.

4. NASA has apparently implemented the architecture recommendations of the Zraket report. Is ESDIS going to be able to meet their responsibility for provision of system-wide software that implements the new EOSDIS architecture?

The August 28th ECS demonstration showed that the current ECS software can perform all critical data ingest, archive, management, processing and distribution functions as defined by the EOSDIS community. Furthermore, it showed that the current ECS software can support the at-launch data ingest rates for all DAACs. Although additional performance tuning and development are needed to support the AM-1 and Landsat-7 launches, ESDIS feels that these results show it will be able to fulfill its responsibility for provision of system-wide software that implements the new EOSDIS architecture.

NASA is also working with DOE to support analysis of the reusability of elements of ECS to support the ARM program.

5. The Zraket report noted that a substantial effort is being made within EOSDIS through the Pathfinder program to reprocess data important for global change research. What lessons learned from the NASA Pathfinder program have been incorporated into EOSDIS?

The Pathfinder program resulted in two categories of products: Software and science algorithms developed to reprocess satellite instrument data to higher level products, and the resulting long term global datasets.

Science algorithms have evolved as science analysis techniques, research interests, and technology has moved forward. Pathfinder algorithms have certainly contributed to these factors, as EOSDIS science algorithms become more developed. In addition, Pathfinder algorithms were used in the early ECS prototype work. The prototype development represented a generic, stand-alone processing environment using the design concepts from Pathfinder, as well as other data processing systems. The prototype was demonstrated using real Pathfinder processing algorithms, both as heritage algorithms and as encapsulated within the SDP Toolkit.

Another early ECS study was aimed at analyzing the performance of the SDP Toolkit with the Pathfinder SSM/I precipitation rate algorithm

(FORTRAN 77) obtained from NASA/MSFC. The ability to use toolkit functions in a parallel symmetric multiprocessing environment was also investigated using an SGI CHALLENGE XL with 8 processors.

Pathfinder data production systems were developed to be robust and operable for specifically known, finite datasets. Many of the broad challenges that EOSDIS is required to meet, such as being evolvable, flexible, extensible, interoperable, etc., were not applicable to Pathfinder software development. In addition, although the Pathfinder data production systems provide several lessons learned, the heart of the EOSDIS implementation challenges results from requirements that have never been implemented before. Specifically, EOSDIS is a distributed system, where each site is required to produce, archive, and distribute data in conjunction with other sites; science software is dependent on products from other science teams; the information management system is very sophisticated to satisfy the broad and very diverse user community; a coordinating infrastructure is required; and capacity and throughput requirements on both the push and pull sides of the system are very high.

The resulting Pathfinder datasets have been invaluable to EOSDIS development. In particular, AVHRR data was utilized by the ECS Science and Technology Lab (STL), to assess various processing technologies (DCE, workstation multiprocessors, MPPs, etc.), evaluate and validate hardware architecture for PGS, and share lessons learned with the science community. This was also done in cooperation with the High Performance Computing and Communications (HPCC) Program, to make use of MPP technologies.

Several DAACs were involved in initial pathfinder data processing activities (NSIDC, GSFC, EDC), and nearly all DAACs are at least involved in or planned to be involved in archival and distribution of Pathfinder products. The Pathfinders have been extremely popular, and have pointed out the need for consistent data sets, at the price of sacrificing potential improvements in the data sets which were discovered later in the processing. The DAACs and the PIs also gained experience in developing processes needed to determine if and when data sets should be reprocessed.

6. The Zraket report concluded that, given the complex funding authority and organization control, the overall coordination of EOSDIS activities would be cumbersome and time-consuming. What is the current chain of command for decision making? For funding?

The DAACs provide annual proposals with work priorities and budget requirements to the ESDIS DAAC Systems and Science Operations Manager, Greg Hunolt. Greg Hunolt has a budget for all of the DAACs (except for ASF, which is managed separately), and has the authority to allocate funds in response to the DAAC proposals. If a good case exists for increased funding, he can provide justification for use of contingency funds to the ESDIS Project Manager. Once their budget submissions are approved, the DAACs have authority to allocate funding to accomplish the proposed work plans.

What are the responsibilities of ESDIS to NASA HQ? Of NASA HQ to ESDIS?

The ESDIS Project is responsible to the MTPE Program Office at Goddard and ultimately to NASA HQ to meet Level 1 requirements and to support science user needs within budget and schedule commitments. NASA HQ and the MTPE Program are responsible to the ESDIS Project to assure that requirements are consistent with the budget.

7. The Zraket report recommended that products should be designed and controlled in part by the scientific and other "customers" of the system. What has been done to implement this recommendation?

The set of standard products is established by the science community. The product generation algorithms and software are developed and delivered by the science community. The priorities for resource allocation for processing and reprocessing the products are governed by the science community through the Data Processing Resource Board (DPRB) chaired by the EOSDIS Project Scientist. The instrument teams and interdisciplinary teams have their own computing facilities where they produce new innovative products (research or special products) that may eventually become standard products. In response to the recommendations from the BSD/CGCR [Board on Sustainable Development/Committee on Global Change Research] from the summer of 1995, the MTPE has further expanded the role of the science community (and others for extensions to applications domain) in generation of products through the WP [Working Prototype]-ESIPs, the proposals for which are now under evaluation. These are meant to provide an extra measure of innovation in both science and technology.

In addition, some DAACs have or are planning to generate additional products as recommended by their UWGs which represent the general user community. This would fall within the scope of DAAC-unique functions the DAACs are expected to develop.

Appendix D

User Survey

The National Research Council's Committee on Geophysical and Environmental Data (CGED) is conducting a review of NASA's Distributed Active Archive Centers (DAACs), each of which manages a different kind of scientific data and serves a unique blend of user communities. The criteria for review focus on how well the DAACs are serving their current scientific user communities and how well positioned they are to serve the much larger community that will use Earth Observing System data and information. Such reviews are beneficial to data centers because they provide critical user feedback to the center and they educate the broader user community about the center's activities. In this sense, the CGED is serving a purpose similar to that of a visiting committee for an academic department.

The review seeks to look at all aspects of the DAACs. In addition to site visits at all of the DAACs, the committee is meeting with representatives from ESDIS, NASA Headquarters, and the ECS contractor to learn about their interactions with the DAACs. Feedback from EOS investigators (both users and producers of data and information) is a critical part of the review, and we ask that you complete and return the following brief questionnaire:

Questionnaire:

1a. Are you a user of the DAACs?
( ) yes
( ) no

1b. If you answered yes to Question 1a, how do you interface with the DAAC?
( ) Web
( ) phone
( ) fax
( ) hard copy
( ) other: Please explain

1c. If you answered yes to Question 1a, are you:
( ) a data provider?
( ) exclusively a data user?
( ) both

2. Which DAAC(s) do you use?
( ) GSFC
( ) LaRC
( ) EDC
( ) ASF
( ) JPL (PO.DAAC)
( ) NSIDC
( ) ONRL

3. How often do you get data?
( ) daily
( ) weekly
( ) monthly
( ) yearly

4. Typical size of data set?
( ) <10 MB
( ) 10-100 MB
( ) 100 MB-1 GB
( ) 1-10 GB
( ) 10 GB

5. How do you get data?
( ) electronic transfer
( ) media

6. Ease in obtaining data in the form you want:
( ) very easy
( ) somewhat easy
( ) average

( ) somewhat difficult
( ) very difficult

7. When problems arise, what do you do?
   ( ) contact friend
   ( ) contact help desk
   ( ) contact DAAC management
   ( ) other. Explain:

8. Overall satisfaction with the DAAC:
   ( ) excellent
   ( ) good
   ( ) average
   ( ) below average
   ( ) poor

9. Major kudos: Use the space below to list major kudos you would give the DAACs

10. Major criticisms: Use the space below to list major criticisms you would level against the DAACs

11. Have you authored any reviewed publications that use DAAC data?
    ( ) yes
    ( ) no
    Provide references (optional)

We appreciate your help.

Anne Linn
National Research Council

The CGED has been reviewing the operations of World and National Data Centers for more than 35 years. Among its most recent reports are *1992 Review of the World Data Center-A for Rockets and Satellites and the National Space Science Data Center (NRC, 1993) and 1993 Review of the World Data Center-A for Meteorology and the National Climatic Data Center* (NRC, 1994.)

TABLE D.1. Summary of Survey Results

User Category	Number of Respondents	Number of GSFC Answers	Number of LaRC Answers	Number of EDC Answers	Number of ASF Answers	Number of JPL Answers	Number of NSIDC Answers	Number of ORNL Answers
Sophisticated	163	103	47	32	27	87	39	21
Casual	46	10	0	2	1	39	4	2
Foreign	184	54	11	9	4	157	22	9
Total	393	167	58	43	32	283	65	32

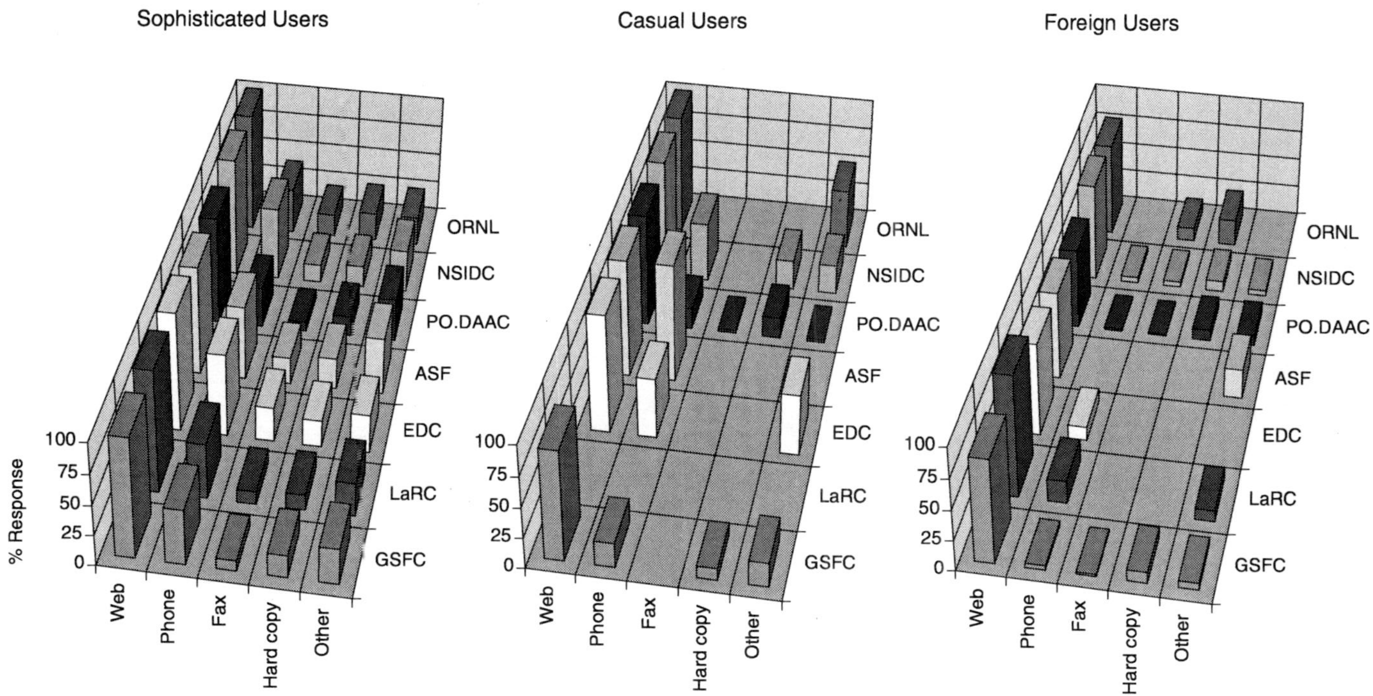

FIGURE D.1. Method for accessing DAACs (see survey, Question 1b). Most respondents chose several access methods, so the total responses for each DAAC is greater than 100%.

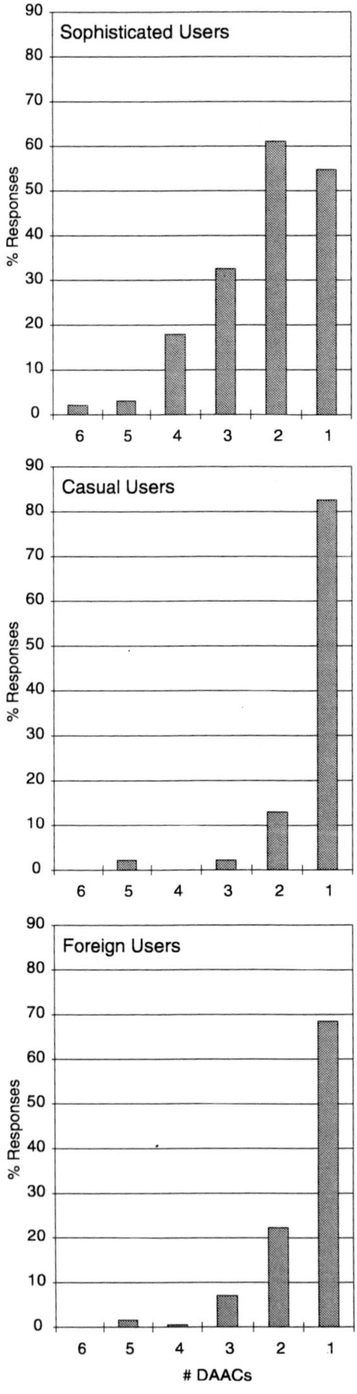

FIGURE D.2. Number of DAAC's used (see survey, Question 3).

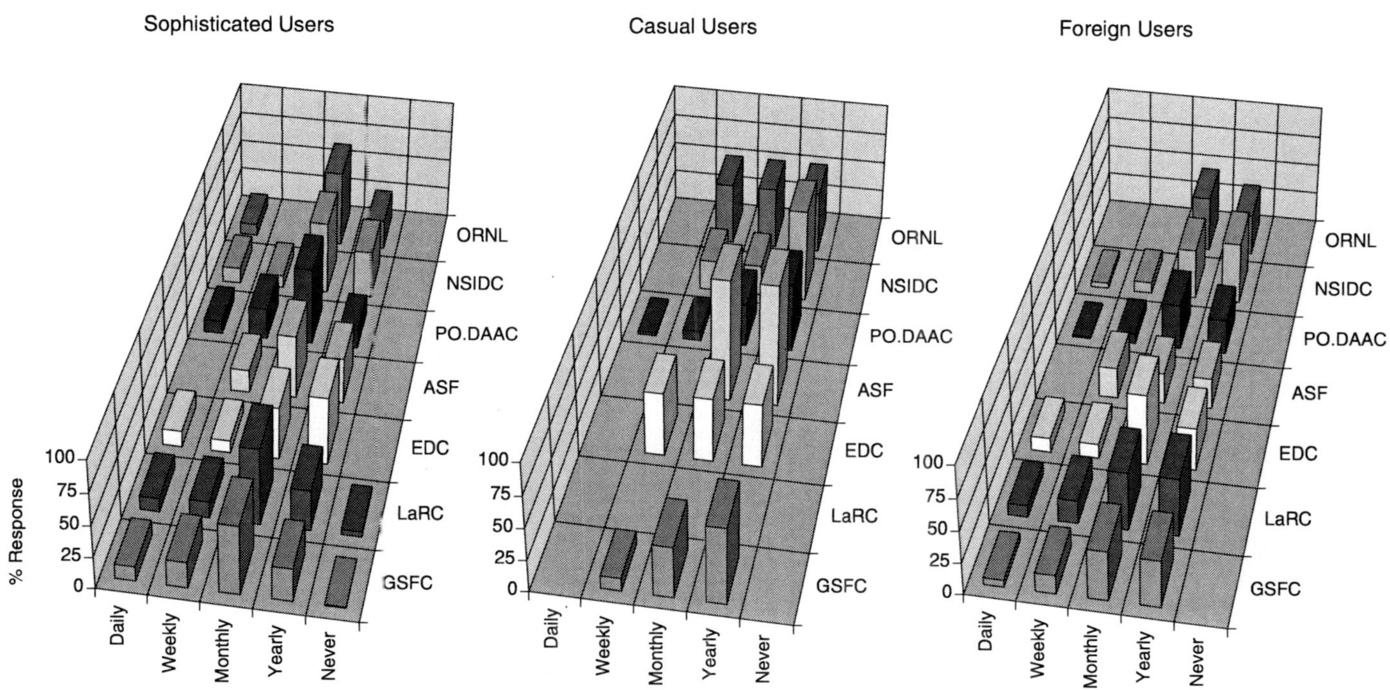

FIGURE D.3. Frequency of DAAC access (see survey, Question 3). Several respondents chose several time intervals, so the total responses for each DAAC may exceed 100%. Sophisticated users who selected "never" are data providers.

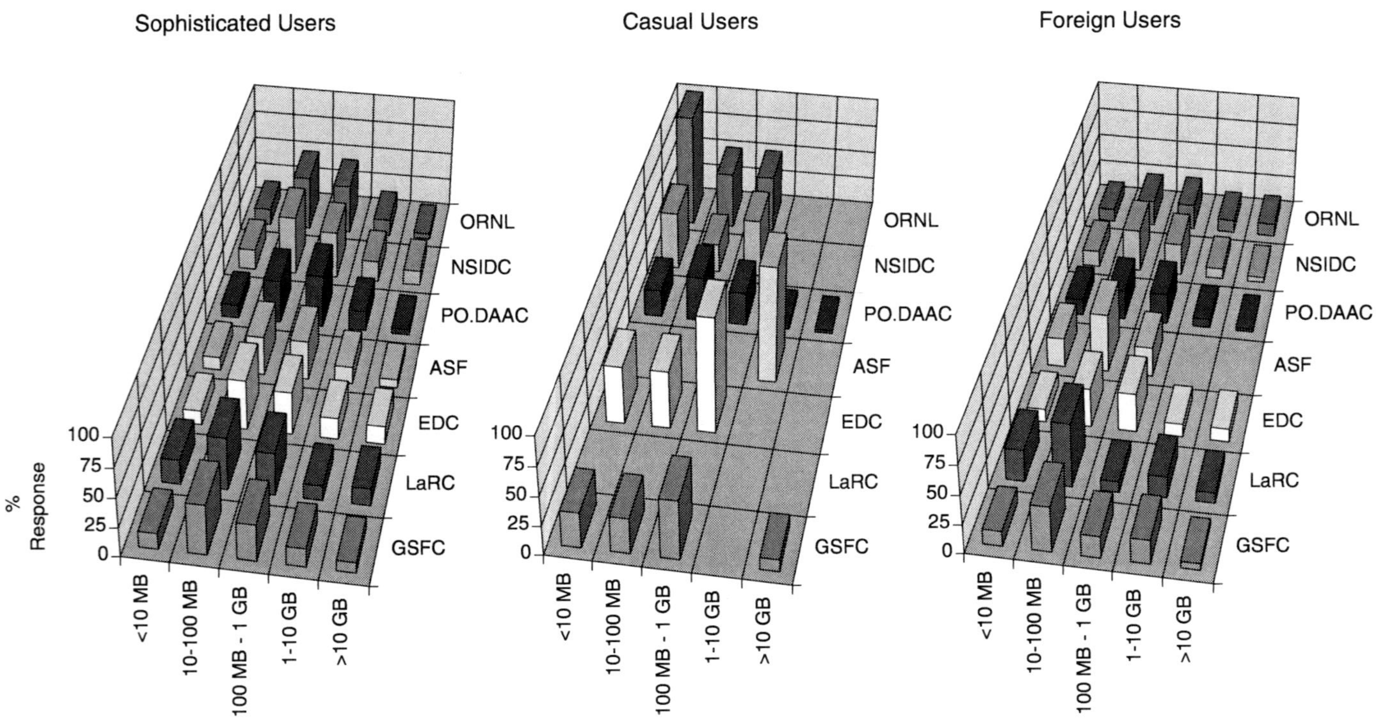

FIGURE D.4. Size of data sets retrieved (see survey, Question 4). Several respondents chose multiple answers, so the total responses for each DAAC may exceed 100%.

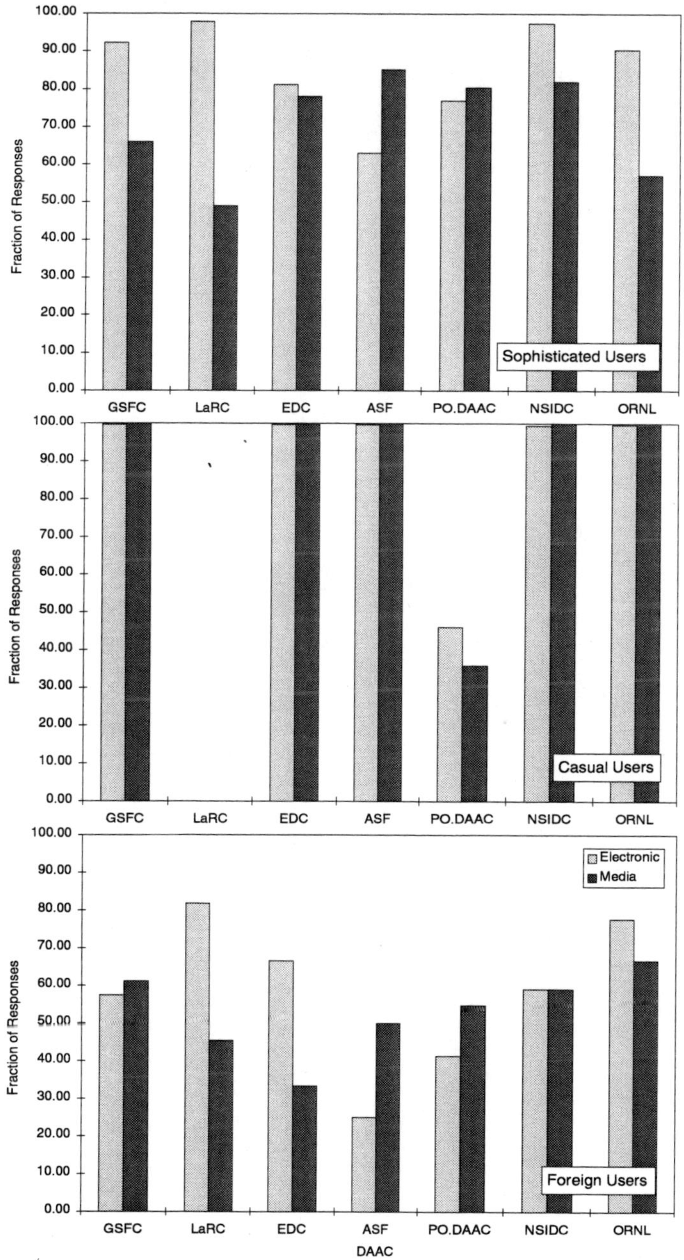

FIGURE D.5. Form of data retrieval (see survey, Question 5). Most respondents chose multiple answers, so the total responses for each DAAC generally exceed 100%.

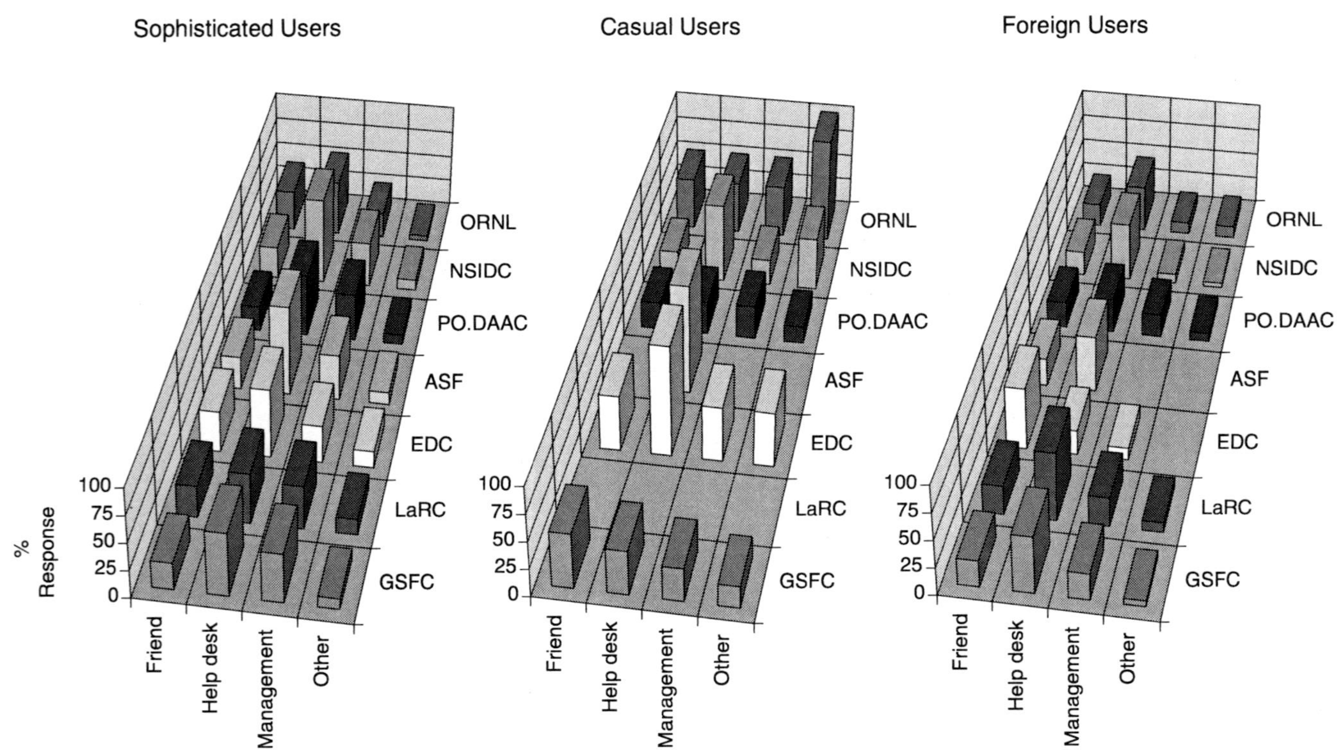

FIGURE D.6. Source of help (see survey, Question 7). Several survey respondents chose multiple means of getting problems addressed, so the total responses for each DAAC commonly exceed 100%.

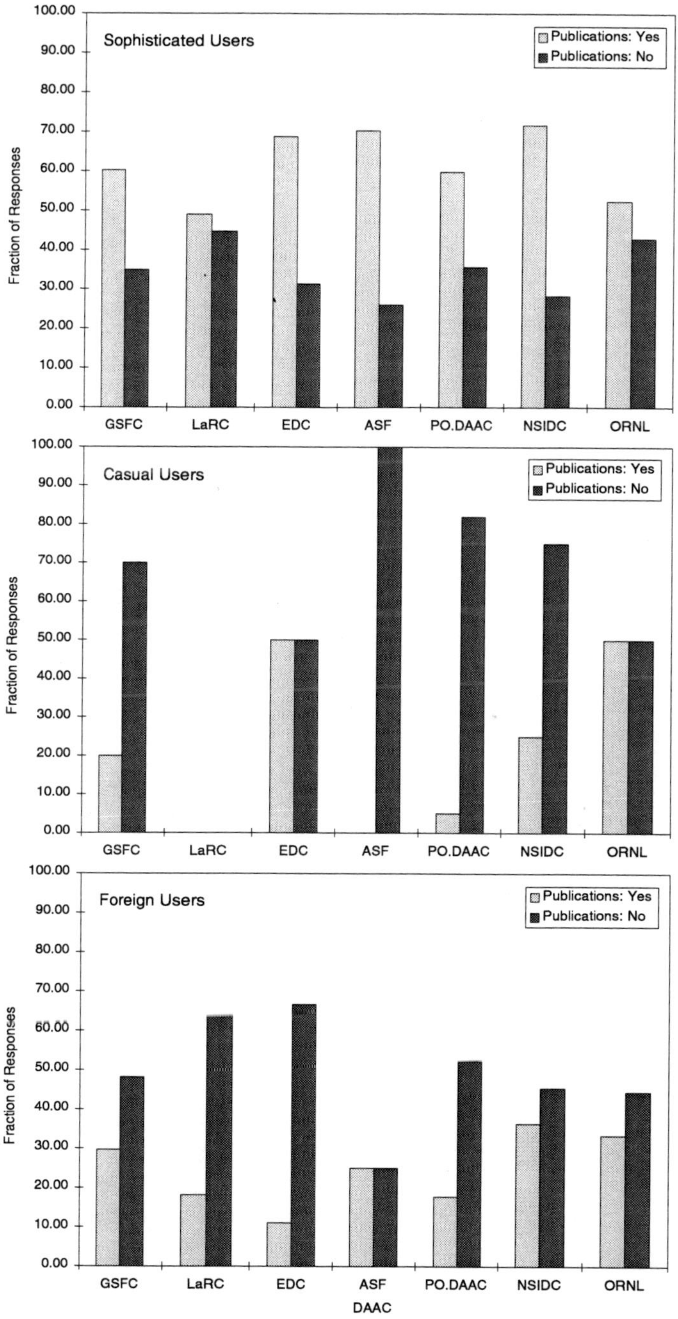

FIGURE D.7. Reviewed publications with DAAC data (see survey, Question 11).

REPRESENTATIVE COMMENTS

GSFC DAAC

- DAAC management is more receptive to the scientists. There is less focus on glitzy stuff, like visualization, and more focus on providing access to data.
- Goddard DAAC certainly listens to users and wants to provide services that are user responsive. We use their interdisciplinary data sets not just for research but also for education (at the doctoral level).
- Whenever I have problems or questions the people at the help desk are very helpful and quick with their responses.
- When a major data revision occurred with the Seawifs data, no attempt was made to say "X file replaces the Y file you already have." We have to run a list of each tape that we have and compare with the listing of the new tape. This process is very time consuming.
- Very helpful in preparing for TRMM data, helping TRMM get visibility.
- Some of their products are hard to get into a format I can use.

LaRC DAAC

- The LaRC DAAC Web site is difficult to navigate and find the necessary information and/or data.
- When I've had difficulty obtaining the data I ordered, some of the user services staff have been extremely helpful.
- There is no way to create a standing order of a data set. If I want all the data from a project, I must make individual orders. For example, ISCCP D1 has monthly data for at least 12 years; that's more than 144 orders.
- They are striving to provide close attention to the requirements for our flight project, and make their most senior technical and managerial staff directly accessible to us.

EDC DAAC

- EDC has been *very* helpful in providing AVHRR and GLCC data sets to us, and has provided great customer service help to us.
- The speed of delivery of Landsat raw imagery is slow, and it is hard to obtain Level 0 data.
- There is evidence that the ordering systems are highly nonautomated. In fact I strongly suspect that at times they keep track of orders using lists created by pen and paper. When one is dealing with a major order, this means that at times they have no idea of even where they are in terms of fulfilling the orders. Sometimes we know better than they do, because their order-tracking system is so disaggregated and incompetent.

- Sometimes overloaded and the response is slow in producing the routine products in large volume. Example, the NALC data set for the United States has taken nerarly a year to copy to CDs.
- When one DAAC manager was asked about why plans for processing data from one of the EOS instruments were apparently so badly behind, he responded that the ECS software was not likely to be ready on time so there was no need for them to be getting ready. This was about a year ahead of the scheduled launch.

ASF DAAC

- The user interface (V0 IMS) is poorly suited to selecting data for repeat-pass SAR interferometry; these data must be selected according to viewed location, date of data collection, spatial baseline, and temporal baseline. The Web interface I have attempted to use catalogs data by location and date only, making cross-references to a database of baselines very difficult.
- Much of the data present in the archive is not visible on the Web interface. Perhaps other interfaces exist but we can't find them.
- ASF is doing a good job with radar interferometry software.
- The Alaska SAR Facility is an excellent resource for valuable, multitemporal SAR data. These data are making it possible to do significant new science. The data request system is working very well, and we get large amounts of data in a timely matter. ASF personnel also deal with us on a personal level and inform us when they have completed projects that may be of interest to us now or in the future.
- ASF just doesn't seem to be able to deal with users who want data outside their station mask.
- When we order data, every single time the order bounces the first few times. Sometimes we can tell why the order isn't accepted; usually we simply receive cryptic messages stating that the order isn't acceptable. Finally, after phone calls to the DAAC staff, which are returned when the people are in town, we can usually straighten out the order. Even if the order is OK, we often get strange disclaimers such as "the order may not have been placed even if it appears to have been placed correctly." On several occasions, we got multiple copies of data, most likely because we didn't know when the order is placed correctly.
- Getting data processed is excruciatingly slow at ASF. This is due to the processor software, not unwillingness. More thought should have been given earlier to a software fix.
- The ASF data consistently have errors in the formatting, and there are a new source of errors with each data set.

PO.DAAC

- Access via the Web site is well-organized and easy to navigate. Fast file transfer times.
- They have been helpful and flexible when I have dealt with them from my position as a data provider. Their performance on distributing and supporting the ERS-1/2 and NSCAT scatterometer data sets has been outstanding. Their flexibility and knowledge about the details of the data sets and processing (resulting from collocation and tight collaboration with JPL Project Offices and working scientists at JPL) is a hallmark of the PO.DAAC operation.
- Incredible array of products turned over to the real world in record time with good documentation and all by folks with nice manners. I remember the time I griped to Chris about the end-of-line terminators in the draft final TOGA CD, and he actually changed the whole thing. That's service.
- The DAAC is *very* responsive to e-mail questions, data arrive promptly, and when a revision was produced, all the information required was sent before I was fully aware I needed to know about it. Supply of data in CD-ROM is currently, for me at least, a very useful and time saving approach.
- There was a long startup time before systematic data distribution occurred, but I suppose that can be attributed to the new nature of the TOPEX data and the problems associated with turning production over to a contractor.
- The JPL PO.DAAC has done a superb job in keeping up with the Geophysical Data Records (GDRs) and revised GDRs for the TOPEX altimeter ever since the launch in 1992.
- The software provided together with the data for reading and manipulating purposes could improve, certain standards of portability and software quality control should be demanded.
- The databases provide students in my classes an opportunity to see and use data sets in problem assignments. It's a valuable resource in that it is readily available and in a form that easily integrates into programs they can use for analysis (spreadsheet, graphics, statistics, etc.).
- HDF is awkward. Some data sets have much more data than I need. Media formats are not always easy to transport from system to system. I would like to see a policy that every data set comes with a sample program to read, subset, and write an ASCII file of the subset. Subsetting should be by variable, time, space, every *n*th, first *m*th, last *q*th. Such a program should be easily modified by the user. I suggest two versions: FORTRAN and C. At a minimum these should be supported for a current PC operating system and a current Sun operating system.

NSIDC DAAC

- Good homepage, good interface for ordering data.
- NSIDC in particular is very much in touch with user needs. It understands the user's issues. NSIDC's and ASF's (for example) science background allow them to interact intelligently about the data being provided. They are always looking at how to best provide the information.
- User services at NSIDC is excellent! Very helpful, promptly address any questions or concerns. They have a lot of "in-house" expertise which I think is essential for a data center—that is, a DAAC must be more than just a "data warehouse," it has to be a resource center for the user community and a focus for acquiring data sets, possibly extracting higher-order data products. I think NSIDC excels at this.
- Communication with the user community is excellent through publications, the Web site, and the active involvement of NSIDC scientists and staff in research activities (the latter is very important). For example, NSIDC staff have attended all of our recent CRYSYS IDS annual meetings and made important contributions to the project.
- Need to have good visualization software for the data. They don't have a lot of imagination when it comes to developing methods to work with the data.

ORNL DAAC

Only one respondent provided written comments:

- DAACs have made strides providing value-added products such as ORNL's NPP products. These products show that good science and good data management are not mutually exclusive.

Acronyms

ACRIM	Active Cavity Radiometer Irradiance Monitor
AMSR	Advanced Microwave Scanning Radiometer
API	application program interface
ARM	Atmospheric Radiation Measurement (DOE archive)
ASF	Alaska SAR Facility
ASP	Alaska SAR Processor
ASTER	Advanced Spaceborne Thermal Emission and Reflection Radiometer
AVHRR	Advanced Very High-Resolution Radiometer
BOREAS	Boreal Ecosystem-Atmosphere Study
CERES	Clouds and the Earth's Radiant Energy System
CGED	Committee on Geophysical and Environmental Data (NRC)
CIESIN	Consortium for International Earth Science Information Networks
CIRES	Cooperative Institute for Research in Environmental Sciences (University of Colorado)
COTS	commercial off-the-shelf
CZCS	Coastal Zone Color Scanner
DAAC	Distributed Active Archive Center
DAO	Data Assimilation Office (GSFC)
DEM	digital elevation model
DLT	digital linear tape
DMSP	Defense Meteorological Satellite Program
DOE	Department of Energy

DVD	digital video disk
ECS	EOSDIS Core System
EDC	EROS Data Center
EOS	Earth Observing System
EOSDIS	EOS Data and Information System
EPA	Environmental Protection Agency
ERBE	Earth Radiation Budget Experiment
EROS	Earth Resources Observations Systems
ERS-1,2	European Remote Sensing Satellites 1 and 2
ESA	European Space Agency
ESDIS	Earth Science Data and Information System (Project Office)
ESE	Earth Science Enterprise (formerly Mission to Planet Earth)
ESIP	Earth Science Information Partner
FIFE	First International Satellite Land Surface Climatology Project Field Experiment
FIRE	First ISCCP Regional Experiment
FTE	full time equivalent
ftp	file transfer protocol
GCMD	Global Change Master Directory
GSFC	Goddard Space Flight Center
GTOPO30	Global 30 Arc-Second Elevation Data Set
HDF	Hierarchical Data Format
IMS	Information Management System
ISCCP	International Satellite Cloud Climatology Project
IWG	Investigators Working Group (EOS)
JERS-1	Japanese Earth Remote-Sensing Satellite
JPL	Jet Propulsion Laboratory
LaRC	Langley Research Center
LaTIS	Langley TRMM Information System
LBA	Large-Scale Biosphere Atmosphere Experiment in Amazonia
MISR	Multi-angle Imaging Spectroradiometer
MODIS	Moderate Resolution Imaging Spectroradiometer
MOPITT	Measurements of Pollution in the Troposphere
MOU	Memorandum of Understanding
MTPE	Mission to Planet Earth (renamed Earth Science Enterprise)
NASA	National Aeronautics and Space Administration
NCSA	National Center for Supercomputing Applications
NGDC	National Geophysical Data Center
NOAA	National Oceanic and Atmospheric Administration
NPP	Net Primary Production
NRC	National Research Council
NSCAT	NASA Scatterometer
NSF	National Science Foundation

NSIDC	National Snow and Ice Data Center
ORNL	Oak Ridge National Laboratory
PGE	Product Generation Executable
PO.DAAC	Physical Oceanography DAAC (JPL)
PoDAG	Polar DAAC User Working Group (NSIDC)
SAGE	Stratospheric Aerosol and Gas Experiment
SAP	Science Advisory Panel
SAR	synthetic aperture radar
SeaSat	Sea Satellite
SeaWiFS	Sea-Viewing Wide-Field-of-View Sensor
SEDAC	Socio-Economic DAAC (CIESIN)
SIR-C	Spaceborne Imaging Radar-C
SMMR	Scanning Multichannel Microwave Radiometer
SSM/I	Special Sensor Microwave/Imager
TOMS	Total Ozone Mapping Spectrometer
TOPEX/Poseidon	Ocean Topography Experiment
TOVS	Television and Infrared Observation Satellite Operational Vertical Sounder
TRMM	Tropical Rainfall Measuring Mission
USGS	U.S. Geological Survey
USGCRP	U.S. Global Change Research Program
UWG	User Working Group
VIRS	Visible Infrared Scanner
WDC	World Data Center
WWW	World Wide Web